CAMBRIDGE MONOGRAPHS
ON MATHEMATICAL PHYSICS

General Editors: W. M. McCrea, D. W. Sciama, J. C. Polkinghorne

GAUGE THEORIES OF WEAK
INTERACTIONS

TO MARY

GAUGE THEORIES OF WEAK
INTERACTIONS

J. C. TAYLOR

Reader in Theoretical Physics, University of Oxford

CAMBRIDGE UNIVERSITY PRESS

CAMBRIDGE

LONDON   NEW YORK   NEW ROCHELLE

MELBOURNE   SYDNEY

CAMBRIDGE UNIVERSITY PRESS
Cambridge, New York, Melbourne, Madrid, Cape Town, Singapore,
São Paulo, Delhi, Dubai, Tokyo, Mexico City

Cambridge University Press
The Edinburgh Building, Cambridge CB2 8RU, UK

Published in the United States of America by Cambridge University Press, New York

www.cambridge.org
Information on this title: www.cambridge.org/9780521295185

First published 1976
First paperback edition 1978
Reprinted 1979
Re-issued 2010

A catalogue record for this publication is available from the British Library

Library of Congress Cataloguing in Publication data

Taylor, John Clayton.
Gauge theories of weak interactions.

(Cambridge monographs on mathematical physics)

Includes bibliographical references and index.
1. Weak interactions (Nuclear physics)
2. Gauge invariance. I. Title

QC794.8.W4T38   539.7.54   75–9092

ISBN 978-0-521-20896-3 Hardback
ISBN 978-0-521-29518-5 Paperback

Foreword to the 1978 Reprint

In this foreword I propose, first of all, to mention some of the developments that have taken place in gauge-theory physics during the four years since this book was written.

New particles

Important predictions of the charm model (see §9.6) have been fulfilled by the discovery of the charmed mesonic resonances D^0 (1863 MeV), $D^\pm$ (1874), D^{*0} (2006), $D^{*\pm}$ (2008) and probably also the charmed strange mesons F (2030) and F^* (2140). For example the D^+ is found in the $K^-\pi^+\pi^+$ channel, but not in the $K^+\pi^-\pi^-$ channel, precisely as expected of a cn bound state with the c-quark decaying to λ via the charged current in equation (9.11).

The notion that the narrow ψ (3095) and ψ' (3684) resonances, and other nearby states, are cc-bound states is now widely accepted, and the spectroscopy is at least qualitatively understood. Thus the four-quark model described in chapter 9 seems to work.

It is probably not the whole truth, however. A heavy lepton τ at 1.9 GeV has been discovered [M. L. Perl et al. (1976) Phys. Lett. 63B, 466]. Also particles (called upsilon) have been found in the 9.5 GeV region with a $\mu^+\mu^-$ decay channel [S. W. Herb et al. (1977) Phys. Rev. Lett. 39 252]. It is not unlikely that these are bound states of a fifth, heavier quark and its anti-particle, in analogy to ψ and ψ'. Given that there are more than four leptons, it is theoretically satisfying to find more than four quarks also, since this helps to maintain the cancellation of γ_5-anomalies (see §13.5). These particles require the existence of at least one new lepton doublet and one new quark doublet representation of the $SU(2)U(\times 1)$ group.

Neutral currents

The experimental situation with regard to non-charge-exchange neutrino scattering has not greatly changed (though, at the time of writing, there is some uncertainty about the value of the cross-section for the reaction (8.35)). All experiments are consistent with the Salam–Weinberg model with a value of $\sin^2\theta_w$ between about 0.2 and 0.3 (see chapters 8 and 9).

Parity violation has been found by the detection of asymmetry in the inelastic scattering of longitudinally polarized electrons on deuterium and hydrogen [C. Y. Prescott *et al.* (1968), *Phys. Lett.* **77B**, 347]. The results well confirm the existence of the neutral-current interaction predicted by the Weinberg–Salam model. Experiments have been done with atoms to search for parity-violating effects caused by the neutral-current interaction (see §9.4). Preliminary reports from some experiments [R. Conti *et al.* (1979), *Phys. Rev. Lett.* **42**, 343] are consistent with the Weinberg–Salam model, but earlier reports from other experiments [L. L. Lewis *et al.* (1977), *Phys. Rev. Lett.* **39**, 795; P. E. G. Baird *et al.* (1977), *Phys. Rev. Lett.* **39**, 798] seemed to show disagreement with the model.

The mass of the Higgs particle

A theoretical argument due independently to S. Weinberg and A. D. Linde [S. Weinberg (1976) *Phys. Rev. Lett.* **36**, 294; A. D. Linde (1976) *JETP Letters* **23**, 73; see also J. Ellis, M. K. Gaillard and D. V. Nanopoulos (1976) *Nucl. Phys.* **B106**, 73] places a useful lower bound on the mass of the Higgs particle (see §9.5). The argument is that, if λ in (8.29) and (8.33) is made too small, the one-loop contributions to the effective potential U in (15.10) become relatively important, and they make $U(f) > U(0)$. The Higgs vacuum state would then not be stable. In general, this argument leads to a lower limit of order $e^2 f$ or $e^2 G_w^{-\frac{1}{2}}$, that is a few GeV. For the Weinberg–Salam model with the observed value of θ_w, the lower limit is about 5 GeV.

Gauge theories of strong interactions

The idea that the forces between quarks are due to an unbroken local $SU(3)_{\mathrm{COLOUR}}$ gauge-invariance ('quantum chromodynamics' or 'QCD') (see §18.4 and 18.5) has received much attention.

On the infra-red divergence question (see §4.4), it has now been shown in many cases that cancellation between real and virtual infra-red divergences, term-by-term in perturbation theory, would take place in massless non-abelian theories just as it does in quantum electrodynamics. Attempts to go beyond perturbation theory are hindered by the fact that the relevant coupling-constant in the infra-red region is unknown but presumably large in asymptotically free theories (§18.5).

A promising way to try to test QCD by perturbative calculations is based upon the fact that, in for example e^+e^- annihilation at high energy, jets of hadrons are seen which seem to come from the decay of virtual quarks. The production of virtual quarks and gluons (the latter being a name for the quanta of the Yang–Mills colour field) can be calculated by ordinary perturbation theory, since the effective coupling strength is small in the high-energy region. It is then hoped that the observed distribution of jets can be inferred from the theoretical distribution of virtual quarks and gluons in a way which is not too dependent upon the unknown low-energy confinement processes by which the decay products transform into actual hadrons [G. Sterman and S. Weinberg (1977) *Phys. Rev. Lett.* **39**, 1436]. The detection of a 'gluon jet' would be very interesting.

Because of asymptotic freedom, the renormalization group (see §18.5) provides a method of making some reliable predictions in QCD about high-energy deep-inelastic lepton scattering. Some analyses of the data [H. L. Anderson *et al.* (1978), *Phys. Rev. Lett.* **40**, 1061; J. G. H. de Groot *et al.* (1979), *Phys. Lett.* **82**b, 292] seem to be consistent with these predictions. Several theorists [see, for example, R. K. Ellis *et al.* (1979), *Nucl. Phys.* **B152**, 285] have argued that the scope of this type of prediction by 'improved perturbation-theory' may be widened to include many high-energy, high-momentum-transfer processes. The general idea is to factorize the cross-section into a 'hard' part, which is calculable by perturbation theory with a small effective coupling-constant, and a structure-function which is not so calculable but which is common to different processes.

On the non-perturbative side, there has been important progress in the study of exact classical solutions of the Yang–Mills equations. There are no static, finite-energy solutions of the pure Yang–Mills equations like the monopole solution of the Higgs model (§7.4), but (in the $SU(2)$ case) solutions have been found of the field equations continued to a 4-dimensional Euclidean space by making time imaginary. [A. A. Belavin, A. M. Polyakov, A. S. Schwartz and Y. S. Tyupkin (1975) *Phys. Lett.* **59**B, 85]. Such classical solutions are interpreted as being relevant to a semi-classical treatment of tunnelling through a potential barrier (where the kinetic energy is negative and so the velocity is imaginary). In the present case, the the Yang–Mills potentials at $it \to \pm \infty$ are pure gauge, i.e. $W_\mu^\alpha \, \tau^\alpha = \Omega^{-1}\partial_\mu\Omega$; but the gauge functions $\Omega(x)$ at $\pm \infty$ are not continously deformable into one another by gauge transformations which reduce to the identity at spatial infinity. (This statement is made precise if the gauge is restricted so that $W_0^\alpha = 0$, and then only a time-independent gauge freedom remains). Thus the tunnelling is interpreted as taking place between topologically distinguished states, and the true vacuum is in general a linear combination of such states [C. G. Callan, R. F. Dashen and D. J. Gross (1976) *Phys. Lett.* **63**B, 334; R. Jackiw and C. Rebbi (1976) *Phys. Rev. Lett.* **37**, 172]. This implies that non-abelian gauge theories have an unexpected richness. In weak interactions, certain conservation laws (like CP and lepton conservation) could be violated by the exponentially small (e^{-137}) tunnelling amplitudes [G. 't Hooft (1976) *Phys. Rev.* **D14**, 3432].

I close this foreword with a comment on the references in this book. They were intended to be a guide to useful reading for English-speaking readers, but not to embody a history of the subject. For example, the important roles played by physicists from Belgium, Holland and the Soviet Union are not adequately reflected in the references.

J. C. TAYLOR

May 1978

Contents

Sections marked * can be omitted without seriously spoiling
understanding of subsequent chapters

Preface

Gauge theories of the weak interactions of elementary particles were proposed in 1967. They began to be properly understood mathematically in 1971, and they had received a modicum of experimental verification by 1974. These theories promise to provide a firm basis from which to understand the weak and electromagnetic forces in nature.

In this book, I have attempted to explain the well-established principles of gauge theories. Elaborate models and the details of present experimental tests are less likely to retain their interest. In the first nine chapters, I have tried to keep physical ideas to the fore. More formal work begins in chapter 10.

I hope the book can be understood by someone acquainted with the basics of quantum field theory and Feynman graphs. More specialized ideas, like gauge-invariance and path-integrals, are explained in the text.

Sections marked * in the contents can be omitted without seriously spoiling the understanding of subsequent chapters.

I am indebted to a great many people for the ideas they have imparted to me in conversation. I will not try to name them all, but I have most often sought advice from Mr J. A. Dixon, Dr D. R. T. Jones and Dr D. A. Ross.

I am very grateful to Mrs E. F. Thomas for preparing the typescript with so much skill and patience.

I am grateful to the author, Professor S. L. Adler, and to the American Institute of Physics for permission to reproduce part of Fig. 1 from *Physical Review*, D **9**, 229–30, no. 1 (1974).

I am very grateful to P. H. Frampton, D. Knight, J. E. Paton and J. C. Polkinghorne for their careful and critical reading of the typescript.

Oxford   J. C. Taylor
December 1974

Notation and conventions

$\hbar = c = 1$ (usually)

Bjorken and Drell (1965) is followed as far as possible

$k \cdot x = k_\lambda x^\lambda = k_0 x_0 - \mathbf{k} \cdot \mathbf{x}, \quad \epsilon_{0123} = +1$

$\gamma_0 \gamma_\lambda$ and γ_5 are Hermitian, $\gamma_5 = i\gamma_0 \gamma_1 \gamma_2 \gamma_3$

∂_λ in a derivative coupling gives $-ik_\lambda$ if k is flowing into the vertex

Relativistic normalization of state vectors is used, so that a factor $(2E)^{\frac{1}{2}} (2\pi)^{\frac{3}{2}}$ is included in each one-particle state vector

The step function $\theta(t) = 1$ or 0 according as $t \gtrless 0$

Multiply propagators and vertices by i in Feynman diagrams

Frequently used notation

a, b, c	suffices on general field $\quad \Phi_a$
A_λ	axial current
$\mathscr{A}_\lambda$	electromagnetic potential
B	suffix 'bare' (usually superior)
B_λ	vector field of $U(1)$ group
c	charm quark
C	suffix Cabibbo
D_λ	covariant differential operator
$\mathscr{D}$	path-integration
e	electron, electron field, charge on electron
e_λ	polarization vector for spin-1 particle
$f_{\alpha\beta\gamma}$	structure constants
f	vacuum-expectation-value of a component of ϕ
F	vacuum-expectation-value of ϕ
$F^\alpha, F_5^\alpha, F_L^\alpha, F_R^\alpha$	generators of chiral group
$F_{\lambda\nu}, F_{\lambda\nu}^\alpha$	electromagnetic and Yang–Mills fields tensors
g, g', g^α	coupling constants
$\hat{g}$	dimensionless coupling constant
g^B, g^0	bare coupling constants
G_W	Fermi weak coupling constant, weak group
$G, \bar{G}$	group

G_F	little-group of F in G
G_S	strong group
$i, j, \ldots$	suffices on Higgs field ϕ
I_a^α	inhomogeneous parts of gauge transformation
j_λ	leptonic weak current
J_λ	hadronic weak current
$\mathscr{J}_\lambda$	total weak current
j_λ^{em}	electromagnetic current
k	4-momentum
l	lepton field, number of closed loops in graph
L	Lagrangian
$\mathscr{L}$	Lagrangian density
$\mathscr{L}_G, \mathscr{L}_S$	'gauge-fixing' Lagrangian density, 'spurion' Lagrangian
L	suffix, left-handed
L_λ	left-handed current
m	mass
M	boson mass
n	quark, dimensions of space-time, number of Goldstone bosons
N	neutron, order of Lie group G
$\mathscr{N}$	Nucleon
0	suffix, 'bare' (superior)
p	4-momentum, quark
q	4-momentum, general quark, eigenvalue of operator in quantum mechanics
Q	charge on quark, quantum-mechanical operator
r	dimension of representation ϕ_i
R	renormalized when superior, right-handed when inferior
R_λ	right-handed current
s_λ	source current
S	action
S_{cl}	classical action
T^α	iso-spin operators
T_{ab}^α	homogeneous part of gauge transformation
t^α	Hermitian matrix representation of G
t_λ	arbitrary time-like vector
u	Dirac spinor, source field
U	effective potential function
U_a^α	differential operator in gauge-fixing term

v^α	source field, real matrix representation of G
v	Dirac spinor
V_λ	vector current
V	potential part of Lagrangian
W_λ^α	Yang–Mills field
x	space-time 4-vector, source field, Feynman integration variable, scaling variable in inelastic scattering
X	generating functional for disconnected graphs
y	weak hypercharge, source field, Feynman integration variable, ratio of lepton energies
Y	hypercharge
z	generating functional in particular gauge
Z	generating functional, renormalization constant
Z_λ	field of Z-particle
$\alpha, \beta, \gamma, \ldots$	suffices (usually superior) for regular representation of G
Γ	generating functional for one-particle-irreducible vertices
δ	infinitesimal gauge transformation increment, functional differential
$\Delta, \bar\Delta$	Faddeev–Popov functional
ϵ	small imaginary part of mass, constant in σ-model
$\epsilon_{\lambda\mu\nu\rho}$	numerical tensor
η	spurion field
θ	scattering angle, step function
θ_C	Cabibbo angle
θ_W	Weinberg angle
ζ	anti-commuting quantity
ϕ_i	scalar fields (Goldstone or Higgs)
Φ_a	all fields
λ	ϕ^4 coupling constants, strange quark
$\lambda, \mu, \nu, \ldots$	Lorentz indices
μ	muon, muon field, scale variable in dimensional regularization
ν	neutrino, neutrino field
ψ	fermion field, $\bar\psi = \psi^\dagger\gamma_0$
π	pion field
Π	multiplet containing pion fields
χ	a Higgs field, fermion source
ξ	gauge parameter

τ^a	Pauli matrices for iso-spin
ω	infinitesimal gauge parameter = spurion field
Ω	finite gauge transformation
†	Hermitian conjugate
~	transpose, classical fields (§ 10.4)
^	unit vector, quantities in new basis (§ 6.5)
′	field with zero vacuum-expectation-value, Cabibbo-rotated quantities

Bold-face type represents 3-vectors, either in real space or iso-spin space.

1

Introduction

The laws governing the weak interactions should not be too hard to uncover. The Born approximation is usually a very good one for weak processes, and a large variety of weak reactions are available for observation. Yet a real theory of weak interactions, approaching Maxwell's theory of electromagnetism in completeness, was not found until 1967 (Weinberg 1967, Salam 1968), thirty-three years after Fermi's initial paper.

By 1963 it was reasonably clear what was the structure of the weak Hamiltonian at *low energies* (Feynman and Gell-Mann 1958, Marshak and Sudarshan 1958, Cabibbo 1963). It was a product of vector (or axial vector) currents, and the hadronic part of the vector current was connected by SU(3) with the electromagnetic current. With the normalization of these currents defined by the current algebra of Gell-Mann (1964), the couplings had a universal strength. It was plausible that the interactions were caused by the exchange of charged spin-1 particles (*W*-mesons) which were rather heavy (at least a few GeV/c^2 in mass). Yet the theory was incomplete. Matrix-elements increased with energy in an impossible way, making cross-sections eventually exceed the unitarity limit. And there was no reliable way of doing higher order calculations: the divergent integrals were uncontrollable.

In these respects, a heavy charged spin-1 particle seems quite unlike the massless spin-1 photon. For in quantum electrodynamics with relativistic perturbation and renormalization techniques, one can calculate, at all energies, to arbitrarily high order in the fine-structure constant $e^2/4\pi\hbar c = 1/137$. (This is perhaps not true for an energy E such that $\ln(E/m_e) \simeq 137$, where m_e is the electron mass. But at these literally astronomical energies, of order 10^{56} GeV, gravitation cannot be left out of account.)

Attempts have been made from time to time (for example, Schwinger 1957) to unite electromagnetism with weak interactions. The fundamental couplings can be of equal strength provided that the

[1]

W-mesons are heavy enough. By generalizing the idea of local gauge-invariance, Yang & Mills (1954) and Shaw (1955) had invented a beautiful model which contained three spin-1 particles (two charged and one neutral, say). But two unanswered questions remained: how to give mass to the spin-1 particles other than the photon; and how to make, for heavy charged spin-1 particles, a renormalizable theory (that is a theory with acceptable high-energy behaviour)?

The solution to these puzzles grew out of ideas imported into relativistic physics by Nambu (1960) from the theory of superconductivity. The BCS theory of superconductivity (Bardeen, Cooper and Schrieffer 1957, Anderson 1958, or see for example the text-book Rose-Innes and Rhoderick 1969) has three important features:

(i) The ground state is not an eigenstate of particle number, although the number operator commutes with the Hamiltonian.

(ii) There is a gap between the energy of the ground state and the energy of the lowest excited state.

(iii) A term

$$\lambda^{-2}\mathscr{A}_i^2 \qquad\qquad (1.1)$$

(where $\mathscr{A}_i$ is the magnetic potential) keeps a magnetic field out of the interior of the superconductor (Meissner effect).

Nambu sought a model of *strong* interaction dynamics which incorporated features (i) and (ii). The vacuum did not share the chiral symmetry of the Lagrangian, and the nucleons acquired a mass (the energy gap). Nambu realized, and its inevitability was emphasized by Goldstone (1961), that there must be massless spin-0 bosons, which the chiral operators could create out of the vacuum. These could be identified with the (light but not massless) pions, provided that a small chiral symmetry-breaking term was added to the Lagrangian. A neutral superfluid would indeed exhibit these Goldstone longitudinal excitations, with energies arbitrarily close to the ground state energy.

Only later (Anderson 1963, Higgs 1964a, Englert & Brout 1964) was it realized that, in his model, Nambu had not mimicked all the features of superconductivity. In BCS theory, there are no excitations reaching down to the ground state. The reason is that the Coulomb field turns the would-be Goldstone modes into plasmons, which have a minimum frequency, the plasma frequency, depending on the penetration depth λ in (1.1). Here, then, might be a device for breaking the Yang–Mills symmetry through the vacuum, and giving mass to the charged spin-1 mesons (one must arrange for electromagnetic gauge-invariance to remain unbroken, even by the vacuum). Along these lines, Salam (1969) and Weinberg (1967) constructed a model for the weak inter-actions (of leptons, at least).

Does the vacuum symmetry-breaking destroy the good high-energy behaviour and renormalizability of the Yang–Mills Lagrangian? Both Weinberg and Salam guessed that it does not. But the proof is a somewhat technical matter in the quantization of gauge theories. 't Hooft (1971a), by making a clever choice of gauge, was able to provide the proof.

The Weinberg–Salam model can be extended to include hadronic weak processes, but with some unease where strangeness is concerned. Is the model right? Most of the predictions concern high energies, and are hard to test. One prediction is that there should be weak processes (involving, for example, neutrinos, and showing parity violation) with no charge-exchange ('neutral current' processes). In experiments on neutrino scattering carried out in the last few years at CERN, at the Fermi National Accelerator Laboratory and at the Brookhaven National Laboratory, such processes have been observed at rates consistent (within the substantial errors) with the model. En-couraging as this is, it is not decisive. People had speculated about neutral currents before the Weinberg and Salam gauge model. And, in case neutrino neutral currents had not been observed, variants of the Weinberg–Salam model were at hand which did not require them (though it would have been necessary to invoke undiscovered heavy leptons).

Sometimes in physics an appealing mathematical formalism has to wait some years before the correct physical interpretation is arrived at. It appears that we are now beginning to understand how to use the Yang–Mills Lagrangian in weak interaction physics (possibly also in strong interaction physics, see chapter 18). The ensuing chapters are intended to expound these ideas.

2

Weak interactions and vector mesons

2.1 Weak interactions

For more details of the material of this chapter, see the book by Marshak, Riazuddin and Ryan (1969) or the review by Bailin (1971).

Weak interactions are defined by having a strength comparable with the Fermi β-decay constant

$$G_{\mathrm{W}} = 1.435 \times 10^{-49}\,\mathrm{erg\,cm^3},$$

or

$$G_{\mathrm{W}}/(\hbar c)^3 = 1.165 \times 10^{-5}\,\mathrm{GeV^{-2}}. \tag{2.1}$$

They break the approximate conservation laws of iso-spin, hypercharge, parity and charge conjugation (CP and time-reversal invariance are also broken, possibly by the weak interactions, possibly by a new class of super-weak interactions).

One would expect to be able to use lowest order perturbation theory on the weak couplings, but the dimensional nature of G_{W} in (2.1) is a warning that the interaction is only 'weak' when not too high energies are involved. Take as an example the weak reaction

$$\bar{\nu}_{\mathrm{e}} e^- \to \mu^- \bar{\nu}_{\mu} \tag{2.2}$$

which is closely connected with muon decay. ($\bar{\nu}_{\mathrm{e}}$ and $\bar{\nu}_{\mu}$ are respectively the electron-type and muon-type anti-neutrinos.) Calculated in Born approximation using a direct Fermi coupling between the four fields, the cross-section has the high-energy behaviour (at centre-of-mass energy E)

$$\sigma \sim G_{\mathrm{W}}^2 E^2/(\hbar c)^4. \tag{2.3}$$

From general scattering theory, a total cross-section at centre-of-mass momentum k has the form (neglecting spins which are inessential for the present argument)

$$\sigma = (\hbar^2 \pi/k^2) \sum_l (2l+1)(2 - \alpha_l - \alpha_l^*), \tag{2.4}$$

where the partial-wave amplitude α_l satisfies

$$|\alpha_l| \leqslant 1. \tag{2.5}$$

[4]

For the direct Fermi coupling, $l = 0$ only; and so (2.3) exceeds its maximum possible value for a value of E of the order

$$[G_W/(\hbar c)^3]^{-\frac{1}{2}} \simeq 300\,\text{GeV}. \tag{2.6}$$

Thus the dimensional coupling constant (2.1) causes the Born approximation to fail at centre-of-mass energies of order $300\,\text{GeV}$. It also makes the theory a 'non-renormalizable' one. Let us briefly explain what this means. In relativistic perturbation theory, one generally encounters integrals, over the momenta of intermediate states, which are ultra-violet divergent. If the number of types of divergent graph remains finite as one goes to higher orders, the theory is renormalizable, in the sense that the infinities can be isolated in a limited number of parameters like masses and coupling constants. These are the arbitrary parameters of the theory, and they can be determined only by experiment. A necessary condition for a renormalizable theory is that the coupling constants should be dimensionless or have dimensions of positive powers of energy (with $\hbar = c = 1$). For a coupling like (2.1), with the dimensions of a negative power of energy, higher powers of G_W are associated with worse infinities and more numerous types of divergent integral. An unlimited number of arbitrary constants is required to absorb the infinities, and the theory has little predictive power. Throughout this book, we shall assume that it is sensible to demand of a physical theory that it be renormalizable in ordinary perturbation theory. (More sophisticated attempts have been made to use non-renormalizable theories by summing infinite series of Feynman graphs. No complete theory of this type exists. See, for example, Lehman and Pohlmeyer 1971.)

2.2 W-mesons and their currents

A theoretically attractive improvement on the Fermi theory of weak interactions is the intermediate vector-meson theory. In this, the existence is postulated of charged spin-1 particles, which we shall call W-mesons.

In terms of a complex 4-vector field $W_\lambda(x)$ corresponding to the W-mesons, the Lagrangian density representing the weak interactions is

$$2^{-\frac{1}{2}}g(\mathscr{J}_\lambda^\dagger W^\lambda + \mathscr{J}^\lambda W_\lambda^\dagger), \tag{2.7}$$

where $\mathscr{J}_\lambda$ is a charged current and g a dimensionless coupling constant. Dividing the weak current into leptonic and hadronic parts, we write

$$\mathscr{J}_\lambda = j_\lambda + J_\lambda. \tag{2.8}$$

The leptonic part is

$$j_\lambda^\dagger = \bar{e}\gamma_\lambda \tfrac{1}{2}(1+\gamma_5)\nu_e + \bar{\mu}\gamma_\lambda \tfrac{1}{2}(1+\gamma_5)\nu_\mu, \qquad (2.9)$$

where $e(x)$ is the electron-positron field, ν_e the field of the electron-type neutrino, etc.

Because it is difficult to calculate with the strong interactions, the explicit form of J_λ is not known. According to the Gell-Mann–Cabibbo theory (Gell-Mann 1964, Cabibbo 1963), J_λ is partially and implicitly characterized as follows. Postulate the existence of left- and right-handed octets of currents

$$L_\lambda^\alpha = \tfrac{1}{2}(V_\lambda^\alpha + A_\lambda^\alpha), \quad R_\lambda^\alpha = \tfrac{1}{2}(V_\lambda^\alpha - A_\lambda^\alpha) \quad (\alpha = 1, ..., 8). \qquad (2.10)$$

The charges

$$F_L^\alpha = \int d^3x\, L_0^\alpha(\mathbf{x}, t), \quad F_R^\alpha = \int d^3x\, R_0^\alpha(\mathbf{x}, t), \qquad (2.11)$$

$$F^\alpha = F_L^\alpha + F_R^\alpha, \quad F_5^\alpha = F_L^\alpha - F_R^\alpha,$$

obey Gell-Mann's (1964) algebra of $SU(3)_L \times SU(3)_R$:

$$[F_L^\alpha, F_L^\beta] = if^{\alpha\beta\gamma}F_L^\gamma, \quad [F_R^\alpha, F_R^\beta] = if^{\alpha\beta\gamma}F_R^\gamma. \qquad (2.12)$$

The F^α are generators of the approximate symmetry group $SU(3)$ of strong interactions. They include the iso-spin and hypercharge operators:

$$T^\alpha = F^\alpha \quad (\alpha = 1, 2, 3), \quad Y = \frac{2}{\sqrt{3}} F^8. \qquad (2.13)$$

The hadronic electric current is

$$J_\lambda^{\text{em}} = V_\lambda^3 + 3^{-\frac{1}{2}} V_\lambda^8. \qquad (2.14)$$

In a quark model with a quark triplet (with quantum numbers given in table 2.1)

$$q = \begin{pmatrix} p \\ n \\ \lambda \end{pmatrix}, \qquad (2.15)$$

the currents might simply be

$$\left.\begin{array}{l} L_\lambda^\alpha = \tfrac{1}{2}\bar{q}\gamma_\lambda \lambda^\alpha \tfrac{1}{2}(1+\gamma_5)q, \\ R_\lambda^\alpha = \tfrac{1}{2}\bar{q}\gamma_\lambda \lambda^\alpha \tfrac{1}{2}(1-\gamma_5)q, \end{array}\right\} \qquad (2.16)$$

where the λ_α are Gell-Mann's 3×3 matrices (see Gell-Mann and Ne'eman 1964).

The weak interactions define an $SU(2)_L$ subgroup of

$$SU(3)_L \times SU(3)_R.$$

TABLE 2.1   *Quark quantum numbers*

Quark	Q	Y	T	B
p	$\tfrac{2}{3}$	$\tfrac{1}{3}$	$\tfrac{1}{2}$	$\tfrac{1}{3}$
n	$-\tfrac{1}{3}$	$\tfrac{1}{3}$	$\tfrac{1}{2}$	$\tfrac{1}{3}$
λ	$-\tfrac{1}{3}$	$-\tfrac{2}{3}$	0	$\tfrac{1}{3}$

It is generated by the charges $F_{\mathrm{L}}'^{\alpha}$ ($\alpha = 1, 2, 3$) associated with the currents

$$L_{\lambda}'^{\alpha} = \tfrac{1}{2}\bar{q}'\gamma_{\lambda}\lambda^{\alpha}\tfrac{1}{2}(1+\gamma_5)q', \qquad (2.17)$$

where

$$p' = p, \quad n' = n\cos\theta_{\mathrm{C}} + \lambda\sin\theta_{\mathrm{C}}, \quad \lambda' = -n\sin\theta_{\mathrm{C}} + \lambda\cos\theta_{\mathrm{C}}, \quad (2.18)$$

θ_{C} being Cabibbo's angle and experimentally

$$\sin\theta_{\mathrm{C}} \simeq 0.22. \qquad (2.19)$$

In this notation, the hadronic weak current is

$$J_{\lambda}^{\dagger} = \bar{n}'\gamma_{\lambda}\tfrac{1}{2}(1+\gamma_5)p \qquad (2.20)$$

or

$$J_{\lambda} = L_{\lambda}'^{1} + \mathrm{i}L_{\lambda}'^{2}. \qquad (2.21)$$

Equation (2.20) should be compared with (2.9).

The algebra (2.12) defines the normalization of the hadronic currents in a way which is not dependent on the quark model forms (2.16) and (2.20). In terms of this normalization, (2.8) states that the weak interactions of leptons and of hadrons have a universal strength.

The structure defined by (2.7), (2.9) and (2.21) is consistent with all known properties of weak interactions, except the recently discovered neutral current effects (see chapters 8 and 9) (except, also, the small *CP* violation observed in K^0 decay, whose explanation may lie outside weak interactions – for further discussion see chapter 17). The $\Delta T = \tfrac{1}{2}$ rule in non-leptonic weak decays is not simply explained by (2.21), though it is not inconsistent with it (see §18.6).

2.3   Heavy vector mesons and high-energy behaviour

We turn now from the detailed structure of the current (2.8) to the dynamics of the *W*-mesons. Take as an example neutron decay

$$N \to Pe^{-}\bar{\nu}_{\mathrm{e}}. \qquad (2.22)$$

To lowest order (second) in g, the amplitude is, from (2.7), (2.8) and (2.9),

$$\tfrac{1}{4}g^2\bar{u}_{\mathrm{e}}\gamma_{\lambda}(1+\gamma_5)v_{\nu}\left[\frac{(q^{\lambda}q^{\mu}/M^2) - g^{\lambda\mu}}{q^2 - M^2}\right]\langle P\,\mathrm{out}\,|\,J_{\mu}(0)\,|\,N\,\mathrm{in}\rangle. \qquad (2.23)$$

Here q is the total 4-momentum of the leptons and u_e, v_ν are their Dirac spinor wave-functions. The lepton matrix-element is evaluated explicitly, since we are neglecting electromagnetic corrections (and higher order weak corrections). $|P\text{ out}\rangle$ and $|N\text{ in}\rangle$ are the final and initial hadron states, and the explicit calculation of the matrix-element of J_μ is a problem in strong interaction physics.

The expression in square brackets in (2.23) is the propagator of the virtual W-meson which is exchanged. The term containing q^λ is, by the Dirac equations satisfied by u_e and v_ν, proportional to the electron's mass, m_e. If $M \gg m_e$ and $M^2 \gg |q^2|$, the propagator is well approximated by

$$g^{\lambda\mu}/M^2. \tag{2.24}$$

In this approximation, the interaction is like a direct one, with an effective Fermi coupling constant

$$G_W = 2^{-\frac{1}{2}}g^2/M^2. \tag{2.25}$$

(The factor $2^{-\frac{1}{2}}$ in (2.7) is inserted to be consistent with the usage in Chapter 8, and the factor $2^{\frac{1}{2}}$ between (2.23) and (2.25) is an historical accident resulting from the definition of G_W before parity violation was discovered.)

We must now comment on the propagator

$$(-g^{\lambda\mu} + q^\lambda q^\mu/M^2)(q^2 - M^2 + i\epsilon)^{-1} \tag{2.26}$$

in (2.23), where the small positive number ϵ defines the contour to be taken round the pole in a Feynman integral. According to a fundamental unitarity principle (see, for example, Eden, Landshoff, Olive and Polkinghorne 1966: 110–16), the imaginary part of (2.26),

$$-i\pi(-g^{\lambda\mu} + q^\lambda q^\mu/M^2)\,\delta(q^2 - M^2), \tag{2.27}$$

is interpreted as describing real (as opposed to virtual) particles. The tensor in (2.27) then comes from a sum over the three polarization states of a spin-1 particle,

$$\sum_{\sigma=1}^{3} e_\lambda^{(\sigma)} e_\mu^{(\sigma)*} = -g_{\lambda\mu} + q_\lambda q_\mu/M^2, \tag{2.28}$$

where $e_\lambda^{(\sigma)}$ are three independent polarization vectors, each satisfying

$$q^\lambda e_\lambda^{(\sigma)} = 0 \quad (\sigma = 1, 2, 3). \tag{2.29}$$

The term $q_\lambda q_\mu/M^2$ is in (2.27) because there are just three spin states, with polarization vectors constrained by (2.29). For example, if q is chosen along the time axis, the $e_\lambda^{(\sigma)}$ each have zero time-component, and (2.28) is the projection operator on to the space of vectors with zero time-component.

Note that in (2.27) M^{-2} is the same thing as $(q^2)^{-1}$, but this is not so in (2.26). To replace M^{-2} by $(q^2)^{-1}$ in (2.26) would be to create a pole at $q^2 = 0$ with no physical interpretation.

Having recognized the inevitability of (2.26), we consider its influence on renormalizability. The propagator for a spin-0 particle is

$$(q^2 - M^2 + i\epsilon)^{-1}.$$

For a spin-$\frac{1}{2}$ particle it is

$$(\gamma \cdot q + m)(q^2 - m^2 + i\epsilon)^{-1}.$$

For a photon, it may be chosen to be

$$-g_{\lambda\mu}(q^2 + i\epsilon)^{-1}.$$

The asymptotic forms of all of these for large values of the components of q are obtained by leaving out the masses. They contain no dimensional parameter, but just the variable q itself. This is why (as mentioned in §2.1) the dimension of the coupling constant determines whether a theory is renormalizable – the above propagators contain no dimensional parameter that is relevant.

The asymptotic form of (2.26), however, is

$$q^\lambda q^\mu / (M^2 q^2),$$

containing the dimensional coefficient M^{-2}. Because of this, theories with charged massive spin-1 mesons are not normally renormalizable although the coupling constant g in (2.7) is dimensionless. The reason is simply that the factors of M^{-2} in higher orders of perturbation theory have to be compensated by more and more divergent integrals.

The theory would be renormalizable only if the offending $q^\lambda q^\mu$ term in (2.26) could be shown to be zero or to be capable of being transformed away in all circumstances. It has been known for some time (see for example Salam 1960) that this is the case for *neutral* massive spin-1 mesons coupled to a divergenceless current. The chief aim of the gauge theories to be described in this book is to generate more complicated models in which the same thing happens for *charged* mesons. The straightforward theory of charged spin-1 mesons, given by (2.7), (2.26), and the electromagnetic interactions of the mesons, certainly does *not* have this property. This has been shown in careful calculations (Veltman 1968, Boulware 1970). We shall be content to take the high-energy behaviour of Born terms as suggestive evidence for or against renormalizability, as we did in § 2.1.

Consider first the amplitude for the lepton reaction (2.2), which is given by an expression similar to (2.23). The main effect of the W-meson is to replace G_W by

$$2^{-\frac{5}{2}}g^2/(M^2-q^2) \qquad (2.30)$$

(the $q^\lambda q^\mu$ terms give unimportant products of lepton masses). Therefore (2.3) is replaced by

$$\sigma \sim g^4 E^2/(M^2-q^2)^2. \qquad (2.31)$$

If E and θ are the energy and scattering angle in the centre-of-mass,

$$q^2 \sim -E^2(1-\cos\theta)$$

for large E; so (2.31) corresponds to a scattering amplitude

$$f(\theta) \sim g^2[1-\cos\theta+(M^2/E^2)]^{-1}. \qquad (2.32)$$

This leads to partial-wave amplitudes which are bounded, except for terms behaving as

$$\ln(E/M). \qquad (2.33)$$

Such terms do indeed violate unitarity at sufficiently high energy, but in a much milder way than (2.3). Perturbation theory certainly breaks down to some extent. Presumably a suitable sum of Feynman graphs can convert the logarithms into a harmless function. However, this type of behaviour occurs with the exchange of *any* particle. It is not evidence against renormalizability. For that we have to look further.

The reaction

$$e^+e^- \to W^+W^- \qquad (2.34)$$

has a contribution from the graph of fig. 2.1. The amplitude is proportional to

$$g^2\bar{u}'\gamma\cdot e'(1+\gamma_5)\gamma\cdot q\gamma\cdot ev/q^2, \qquad (2.35)$$

where u', v are the Dirac spinor wave-functions for e^-, e^+, and e, e', are the polarization vectors for W^+, W^-. Take the special case in which W^+ is longitudinally polarized (helicity 0) in, say, the centre-of-mass. Then

$$e = M^{-1}(|\mathbf{k}|, k_0\hat{\mathbf{k}}), \qquad (2.36)$$

since this is invariant under rotations about $\mathbf{k}$, and is normalized and satisfies (2.29). The identity

$$e_\lambda = M^{-1}k_\lambda + \left[\frac{M}{k_0+|\mathbf{k}|}\right](-1,\hat{\mathbf{k}}) \qquad (2.37)$$

is easily verified. Into (2.35) substitute the first term of (2.37). Use the

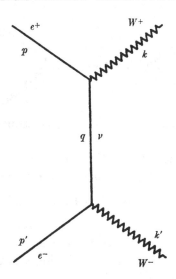

FIGURE 2.1. Feynman graph for weak W-pair production.

Dirac equation for v and neglect the electron mass, to obtain

$$g^2 M^{-1} \bar{u}' \gamma \cdot e' (1 + \gamma_5) v. \tag{2.38}$$

This contains only a few partial waves, and it is easy to verify that their high-energy behaviour contradicts (2.5) – really just because of the dimensional factor M^{-1}. The high-energy behaviour is worse if e' is longitudinal than if it is transverse, but unitarity is contradicted in either case. The second term in (2.37) cannot alter these conclusions.

This example illustrates that it is the M^{-1} factor in (2.36) which is at the root of the violation of unitarity in Born terms, just as it is the M^{-2} term in (2.26) which stops the theory being renormalizable.

There is another mechanism for the reaction (2.34), involving a virtual photon (figure 2.2). It too violates unitarity at high energies (if at least one of the Ws is longitudinal). Could there be a cancellation between the two amplitudes at high energies? (This would relate g to e, but there is no harm in that.)

The electromagnetic couplings of a charged spin-1 particle are not uniquely determined, because its magnetic moment and electric quadrupole moment are not fixed in any simple way. For reasons that will emerge in chapter 4, we here choose the right-hand vertex in fig. 2.2 to be

$$e[(k' - k)_\lambda g_{\mu\nu} + (P - k')_\mu g_{\nu\lambda} + (k - P)_\nu g_{\lambda\mu}], \tag{2.39}$$

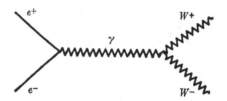

FIGURE 2.2. Feynman graph for electromagnetic W-pair production.

where, to bring out the symmetry, we set $P = -(k+k')$. With the first term of (2.37) in mind, multiply (2.39) by k^μ/M to obtain

$$eM^{-1}[(k'^2 - P^2)g_{\nu\lambda} + P_\nu P_\lambda - k'_\nu k'_\lambda].\qquad(2.40)$$

The last two terms here make no contribution to the amplitude, because of the Dirac equations for $\bar{u}$ and v' and because $k'\cdot e' = 0$. The photon propagator is $(P^2)^{-1}$, and combining this with (2.40) one obtains the high P^2 limit

$$-e^2M^{-1}\bar{u}'\gamma\cdot e'v.\qquad(2.41)$$

This is of almost the right structure to cancel (2.38), except that it lacks the γ_5 term. Even if cancellation were possible in this case, it is not possible for the reaction

$$\nu\bar{\nu} \to W^+W^-,\qquad(2.42)$$

for here there is an electron-exchange graph analogous to fig. 2.1, but no electromagnetic process like fig. 2.2.

It is, then, impossible to achieve acceptable high-energy behaviour with charged W-mesons and photons (and the known leptons) alone. We shall see in chapter 8 how additional particle exchanges can accomplish the cancellation of the badly behaved terms.

3

Photons

In the last chapter we gave the physical reasons why charged massive spin-1 particles are generally associated with bad high-energy behaviour and lack of renormalizability. How is it, then, that the theory of photons, which after all have spin 1, does not suffer from these troubles? The most obvious difference between the two cases is that photons have no mass, and so the bad terms containing M^{-1} in (2.26) and (2.37) cannot possibly be there in quantum electrodynamics. It will transpire that this is very closely connected, though not quite identical, with the essential point, which is gauge-invariance.

3.1 Photon spin states

Because photons are massless, they have only the two transverse polarization states (or helicities ± 1). They cannot be longitudinally polarized. This is ultimately a consequence of Lorentz covariance. Let $J_{\lambda\nu}$ be the quantum-mechanical generators of Lorentz transformations, forming an antisymmetric 4×4 tensor. The spin of a particle of 4-momentum k_μ is represented by the Pauli–Lubanski vector (see, for example, Omnès 1970 chapter 4)

$$w_\lambda = \tfrac{1}{2}\epsilon_{\lambda\mu\nu\rho}k^\mu J^{\nu\rho} \tag{3.1}$$

which has three independent components. The spin states of a particle of non-zero mass m may be characterized by the pair of eigenvalue equations

$$\left.\begin{aligned} [w^2 + m^2 s(s+1)]|s,\sigma\rangle &= 0, \\ (n_\lambda w^\lambda - m\sigma)|s,\sigma\rangle &= 0. \end{aligned}\right\} \tag{3.2}$$

Here s is the magnitude of the spin and σ is its projection along an arbitrary 4-vector n_λ satisfying

$$n^2 = -1, \quad n_\lambda k^\lambda = 0. \tag{3.3}$$

For zero mass, however, one may consistently impose the stronger relations

$$(w_\lambda - \sigma k_\lambda)|\sigma\rangle = 0, \tag{3.4}$$

where σ is an integer or half-integer (the helicity). If $m \neq 0$, (3.4) would be inconsistent, since (3.1) satisfies $k \cdot w \equiv 0$. If parity is a good quantum number, it connects states with opposite signs of σ. For a photon $\sigma = \pm 1$.

In field theory, every particle is associated with a field with definite transformation properties under the homogeneous Lorentz group. For a massive spin-1 particle, there is no difficulty in associating a 4-vector $e^{(\sigma)}(k)$ with a spin state σ. There are just the required number of three independent vectors satisfying

$$e^{(\sigma)*} \cdot e^{(\sigma')} = -\delta_{\sigma\sigma'}, \quad k \cdot e^{(\sigma)} = 0. \tag{3.5}$$

To construct a conventional quantum theory of Maxwell's equations it also proves necessary to use a 4-vector, the electromagnetic 4-potential $\mathscr{A}_\lambda(x)$. (Mandelstam (1968) has shown that it is possible to work with the fields $F_{\mu\nu}(x)$, but at the cost of some complications.) Since there are only two spin states, we require two independent vectors $e_\lambda^{(\sigma)}$. Indeed, for $k^2 = 0$, there are only two independent solutions of (3.5). (The vector (2.36) which represents longitudinal polarization for non-zero mass, has no limit for $M \to 0$.) The trouble is that the polarization vector $e^{(\sigma)}$ is not uniquely determined by the helicity σ. If $e^{(\sigma)}$ is an appropriate vector, so is

$$e_\lambda^{(\sigma)'} = e_\lambda^{(\sigma)} + \omega k_\lambda, \tag{3.6}$$

for any value of ω. This is because $e^{(\sigma)'}$ satisfies (3.5), and because the spin state of the particle is defined by the Lorentz transformations which leave k_λ invariant (these are generated by w_μ); and under these e and e' clearly transform identically.

To make a photon polarization vector unique, one may, for example, choose a vector t_λ and supplement (3.5) with the condition

$$t \cdot e^{(\sigma)} = 0. \tag{3.7}$$

The trouble with this is that it sacrifices Lorentz invariance, since t_λ is chosen arbitrarily. Under a Lorentz transformation

$$e_\lambda \to \Lambda_\lambda^\nu e_\nu + \omega(\Lambda, t) k_\lambda, \tag{3.8}$$

where Λ is the ordinary matrix of the transformation and ω is chosen to restore condition (3.7) (we look upon t_λ as not transforming under the Lorentz transformation). This point of view has been taken by Weinberg (1965a).

3.2 Feynman rules for photons

We can now return to the question: why do photons not suffer from the same troubles as massive spin-1 particles? The bad high-energy behaviour of Born terms came from longitudinal polarization states, and these simply do not exist for photons. The trouble with renormalization came from the propagator (2.26). To find the analogous function for a photon, we can reverse the sequence of equations (2.26), (2.27), (2.28). From (3.5) and (3.7), the analogue of (2.28) is

$$\sum_{\sigma} e_{\lambda}^{(\sigma)} e_{\mu}^{(\sigma)*} \equiv P_{\lambda\mu}(k) = -g_{\lambda\mu} - [t^2 k_{\lambda} k_{\mu} - k \cdot t(k_{\lambda} t_{\mu} + k_{\mu} t_{\lambda})]/(k \cdot t)^2. \quad (3.9)$$

This leads to a propagator, analogous to (2.26),

$$P_{\lambda\mu}(k)\,(k^2 + i\epsilon)^{-1}. \quad (3.10)$$

There is no dimensional parameter (like M^{-2}) in this propagator, and one therefore expects the theory to be renormalizable.

(The propagator (3.10) is only strictly correct for t space-like – the so-called 'axial' gauge. The Coulomb gauge has t time-like, and there the correct propagator is

$$\{-g_{\lambda\mu} - [t^2 k_{\lambda} k_{\mu} - k \cdot t(k_{\lambda} t_{\mu} + k_{\mu} t_{\lambda})]/[(k \cdot t)^2 - t^2 k^2]\}\,(k^2 + i\epsilon)^{-1};$$

the numerator in braces is the same as $P_{\lambda\mu}$ at the pole $k^2 = 0$.)

We are now faced with other questions. Why are physical results unchanged by gauge transformations (3.6) for photons in initial or final states? Why do physical results not depend upon the arbitrary vector t_{λ} in (3.9)? The answer to both questions lies in the Ward–Takahashi identities, which are connected with current conservation. We will treat these identities more generally in chapter 12. For the moment we will be content with a few short remarks.

As usual in quantum field theory (see for example Gasiorowicz 1966: 102) the amplitude $M_{\lambda\nu}$ for a process involving two photons, for example, is proportional to the Fourier transform of

$$\langle A \text{ in} | T(J_{\lambda}^{\text{em}}(x) J_{\nu}^{\text{em}}(y)) | B \text{ out}\rangle, \quad (3.11)$$

where the states A and B contain the other particles besides the two photons, and T is the time-ordering symbol,

$$T(J(x) J(y)) = \tfrac{1}{2}\{J(x), J(y)\} + \tfrac{1}{2}\epsilon(x_0 - y_0) [J(x), J(y)], \quad (3.12)$$

where $\epsilon(t) = \pm 1$ for $t \gtrless 0$. A product like $k^{\lambda} M_{\lambda\nu}$ (where a polarization vector $e_\lambda(k)$ has been replaced by its 4-momentum k_λ) gives rise to a divergence

$$\partial^\lambda T\left(J^{\text{em}}_\lambda(x)\, J^{\text{em}}_\nu(y)\right) = T(\partial^\lambda J^{\text{em}}_\lambda(x)\, J^{\text{em}}_\nu(y)) + \delta(x_0 - y_0)\left[J^{\text{em}}_0(x), J^{\text{em}}_\nu(y)\right].$$
(3.13)

The first term on the right-hand side vanishes by current conservation. The equal time commutator in the second term is the same as for free fields and is zero, at least if the current is constructed from spin-$\frac{1}{2}$ fields. For bosons, there are seagull terms to be added to (3.11) which cancel Schwinger terms in the commutator (Jackiw 1972: 110). These complications do not affect the fundamental physical result

$$k^\lambda M_{\lambda\nu} = 0.$$
(3.14)

We can now examine the apparent t-dependence of (3.9). What we want to show is that (3.9) can be replaced by $-g_{\lambda\nu}$ in applications of the unitarity condition. We choose an example which contains all the essential points. Consider a contribution to the scattering process $\gamma e^- \to \gamma e^-$ in which there are $\gamma\gamma e^-$ intermediate states, as in fig. 3.1. Unitarity demands that the scattering amplitudes should have a cut associated with these intermediate states. The discontinuity across this out is obtained from fig. 3.1. by taking the three internal lines to represent on-shell physical particles. Therefore it has the form

$$M_{\lambda\lambda'}\cdot\delta^+(k^2)\, P^{\lambda\nu}(k)\, \delta^+(k'^2)\, P^{\lambda'\nu'}(k')\, \delta^+(p^2 - m^2)\, (\gamma\cdot p + m)\, N^*_{\nu\nu'},$$ (3.15)

where $P^{\lambda\nu}$ is defined by (3.9) and

$$\delta^+(k^2) = \theta(k_0)\, \delta(k^2).$$
(3.16)

Suffices other than those belonging to the internal photons have been suppressed.

For the amplitudes $M_{\lambda\lambda'}$ and $N_{\nu\nu'}$ in (3.15) we may write expressions like (3.11), where the photons in the initial and final states and the electrons are all included in the state vectors A and B. It is essential to the argument at this stage that the electron is a physical on-shell particle, not a virtual one. In other words, we are using fig. 3.1 as a unitarity diagram not a Feynman diagram. Identities like (3.14) allow both $P_{\lambda\nu}$ and $P_{\lambda'\nu'}$ in (3.15) to be replaced by

$$P_{\lambda\nu} \to -g_{\lambda\nu} + \beta k_\lambda k_\nu,$$
(3.17)

where β is an arbitrary quantity. Finally, the rules for the analytic properties of Feynman diagrams (Eden, Landshoff, Olive and Polking-

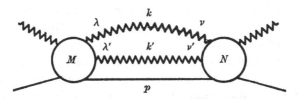

FIGURE 3.1. Unitarity graph.

horne 1966: 110–16) assure us that (3.15) is the imaginary part of a Feynman diagram of the same form as fig. 3.1, in which the photon propagators are

$$(-g_{\lambda\nu} + \beta k_\lambda k_\nu)(k^2 + i\epsilon)^{-1}. \qquad (3.18)$$

Thus, it is consistent with unitarity to use (3.18) in place of (3.10). Two important choices for β in (3.18) are $\beta = 0$, the Feynman gauge, and $\beta = (k^2)^{-1}$, the transverse or Landau gauge.

In a similar way, equations like (3.14) show that, for a photon in an initial or final state, amplitudes are unchanged by (3.6) or by the second term in (3.8).

Why can one not use similar arguments to remove the unwanted $q^\lambda q^\mu / M^2$ term from (2.26)? The reasons are that $\partial \cdot \mathscr{J} \neq 0$ for the weak currents, and, even more fundamentally,

$$\delta(x_0 - y_0)\,[\mathscr{J}_0(x), \mathscr{J}_\nu^\dagger(y)] \neq 0 \qquad (3.19)$$

since the current is charged (as one can see, for instance, from (2.21) and (2.12)); so that the analogue of (3.13) is certainly not zero.

3.3 Gauge-invariance

In this chapter and the preceding one, we have tried to bring out as simply as possible how the high-energy behaviour of Born terms is related to the physical properties of spin-1 particles. We must now be a little more formal, and review the equations of electrodynamics. The Lagrangian density for an electron field $\psi(x)$ interacting with the electromagnetic 4-potential $\mathscr{A}_\lambda(x)$ is

$$\mathscr{L} = -\tfrac{1}{4}(\partial_\lambda \mathscr{A}_\nu - \partial_\nu \mathscr{A}_\lambda)(\partial^\lambda \mathscr{A}^\nu - \partial^\nu \mathscr{A}^\lambda)$$
$$+ i\bar{\psi}(\gamma \cdot \partial - ie\gamma \cdot \mathscr{A})\psi - m\bar{\psi}\psi. \qquad (3.20)$$

This is invariant under the local gauge transformation

$$\left. \begin{aligned} \mathscr{A}_\lambda(x) &\to \mathscr{A}_\lambda(x) + \partial_\lambda \omega(x), \\ \psi(x) &\to \exp\left[ie\omega(x)\right]\psi(x), \end{aligned} \right\} \qquad (3.21)$$

which is related to (3.6). Consequently the equations of motion do not determine $\mathscr{A}_\lambda$ uniquely; and so the propagator, a Green's function of the equation of motion for $\mathscr{A}_\lambda$, does not exist.

In order to quantize the theory, $\mathscr{A}_\lambda(x)$ must somehow be made unique, which requires the gauge-invariance to be broken. This can be done by, for example, adding to (3.20) a gauge-fixing term

$$\mathscr{L}_{\mathrm{G}} = -\tfrac{1}{2}\xi^{-1}(\partial \cdot \mathscr{A})^2. \tag{3.22}$$

The modified equation of motion then is

$$-[g_{\lambda\nu}\Box - (1 - \xi^{-1})\,\partial_\lambda\partial_\nu]\,\mathscr{A}^\nu = e\bar{\psi}\gamma_\lambda\psi. \tag{3.23}$$

The inverse of the differential operator on the left-hand side of this equation gives the propagator to be (in momentum space)

$$[-g_{\lambda\nu} + (1 - \xi)\,k_\lambda k_\nu/k^2]\,(k^2 + i\epsilon)^{-1}, \tag{3.24}$$

which is of the form of (3.18). The Feynman gauge is $\xi = 1$, and the Landau gauge is reached in the limit $\xi \to 0$.

We have not shown that (3.22) can be added to (3.20) without spoiling the physical content of the theory. The argument of §3.3 showed that the adoption of (3.24) for the propagator is consistent with unitarity. But we postpone to chapter 11 a more complete treatment of the quantization of gauge fields.

4

The Yang–Mills field

4.1 The field equations

We have seen that theories of charged spin-1 mesons are in general not renormalizable, but quantum electrodynamics is renormalizable. These facts perhaps suggest that gauge-invariance is a crucial factor in the renormalizability of vector-meson theories. We therefore describe how a locally gauge-invariant theory of charged vector mesons can be constructed (Yang and Mills 1954, Shaw 1955). The theory represents, apparently, massless spin-1 particles only, and so it can have no direct physical application (at least in the context of conventional perturbation theory). We ignore this point for the moment: the contents of this chapter will turn out to be a step towards a physical model.

The basic idea is to generalize the single-parameter group $U(1)$ of local phases $\exp(ie\omega)$ in (3.21) to a more general group G. In particular G will be non-abelian, that is, will contain non-commuting elements. The simplest example to have in mind for G is the group $SU(2)$ like the iso-spin transformations of nuclear physics (or, for that matter, the rotational transformations of spin-$\frac{1}{2}$ states). We will first treat $SU(2)$ as an example, and later generalize to an arbitrary group. Because of its familiarity, we will sometimes use the language of 'iso-spin', not intending thereby to identify the transformations with those of nuclear physics.

Suppose $\psi(x)$ represents an isotopic doublet (spin-$\frac{1}{2}$ representation of $SU(2)$); so that ψ is a two-element column matrix. It is easy to construct a Lagrangian which is invariant under the *global* transformation

$$\psi(x) \rightarrow \exp\left(\tfrac{1}{2}ig\boldsymbol{\tau}\cdot\boldsymbol{\omega}\right)\psi(x), \tag{4.1}$$

in which the exponential is just a parametrization of a general 2×2 unitary unimodular matrix. The components of the vector $\boldsymbol{\tau}$ are the Pauli matrices, and $\boldsymbol{\omega}$ is an arbitrary space and time independent vector. The constant g is factored out at this point to accord with common usage. It will turn out to be a coupling constant.

[19]

To generalize (4.1) to *local* transformations we make $\boldsymbol{\omega}$ a function of the space-time point x, obtaining

$$\psi(x) \to \exp\left[\tfrac{1}{2}ig\boldsymbol{\tau}\cdot\boldsymbol{\omega}(x)\right]\psi(x). \qquad (4.2)$$

A Lagrangian invariant under (4.1) will not in general be invariant under (4.2), because of the presence of derivatives of $\psi(x)$. Invariance can be restored by replacing the gradient ∂_λ by a covariant derivative

$$D_\lambda \equiv \partial_\lambda - \tfrac{1}{2}ig\boldsymbol{\tau}\cdot\mathbf{W}_\lambda, \qquad (4.3a)$$

where $\mathbf{W}_\lambda$ is an isotopic triplet vector field. Then $D_\lambda\psi$ transforms in the same way as ψ under (4.2) provided that

$$\boldsymbol{\tau}\cdot\mathbf{W}_\lambda \to \exp\left(\tfrac{1}{2}ig\boldsymbol{\tau}\cdot\boldsymbol{\omega}\right)\left[\boldsymbol{\tau}\cdot\mathbf{W}_\lambda + 2ig^{-1}\partial_\lambda\right]\exp\left(-\tfrac{1}{2}ig\boldsymbol{\tau}\cdot\boldsymbol{\omega}\right). \qquad (4.3b)$$

For most practical purposes it suffices to restrict $\boldsymbol{\omega}(x)$ to be infinitesimal. Then, to first order, (4.3b) becomes

$$\mathbf{W}_\lambda \to \mathbf{W}_\lambda + \partial_\lambda\boldsymbol{\omega} - g\boldsymbol{\omega}\wedge\mathbf{W}_\lambda. \qquad (4.4)$$

Equation (4.4) makes clear the dual role of the field $\mathbf{W}_\lambda$. The *inhomogeneous* term $\partial_\lambda\boldsymbol{\omega}$ shows that the source of the W-field is iso-spin, just as the electromagnetic potential transforms inhomogeneously under (3.21) and has electric charge for its source. But the *homogeneous* term $-\boldsymbol{\omega}\wedge\mathbf{W}_\lambda$ shows that $\mathbf{W}_\lambda$ is an iso-vector, that is to say its quanta themselves contribute to the iso-spin. Among other things, therefore, $\mathbf{W}_\lambda$ must be coupled to itself.

With the definition (4.3a), the fermion Lagrangian is, in analogy with (3.20),

$$\bar{\psi}(i\gamma\cdot D - m)\psi. \qquad (4.5)$$

In order to construct a Lagrangian for $\mathbf{W}_\lambda$ itself, we have to find an invariant. The first step is to construct an antisymmetric tensor $\mathbf{F}_{\lambda\nu}$, in some respects like the electromagnetic field tensor, but transforming properly as an iso-triplet (with no inhomogeneous term like that in (4.4)). The trick which does this is to define

$$\boldsymbol{\tau}\cdot\mathbf{F}_{\lambda\nu} = 2ig^{-1}[D_\lambda, D_\nu] \qquad (4.6)$$

or

$$\mathbf{F}_{\lambda\nu} = \partial_\lambda\mathbf{W}_\nu - \partial_\nu\mathbf{W}_\lambda + g\mathbf{W}_\lambda\wedge\mathbf{W}_\nu. \qquad (4.7)$$

Then, under (4.4),

$$\mathbf{F}_{\lambda\nu} \to \mathbf{F}_{\lambda\nu} - g\boldsymbol{\omega}\wedge\mathbf{F}_{\lambda\nu}. \qquad (4.8)$$

The appropriate invariant Lagrangian, generalizing the first term of (3.20), is

$$-\tfrac{1}{4}\mathbf{F}_{\lambda\nu}\cdot\mathbf{F}^{\lambda\nu} = -\tfrac{1}{2}(\partial_\lambda\mathbf{W}_\nu - \partial_\nu\mathbf{W}_\lambda)\cdot\partial^\lambda\mathbf{W}^\nu$$
$$-g(\mathbf{W}_\lambda\wedge\mathbf{W}_\nu)\cdot\partial^\lambda\mathbf{W}^\nu - \tfrac{1}{4}g^2[(\mathbf{W}_\lambda\cdot\mathbf{W}^\lambda)^2 - \mathbf{W}_\lambda\cdot\mathbf{W}_\nu\mathbf{W}^\lambda\cdot\mathbf{W}^\nu]. \qquad (4.9)$$

This confirms, as anticipated, that $\mathbf{W}_\lambda$ is self-coupled.

(In many ways it would be more elegant to express everything in terms of a field $\mathbf{W}'_\lambda = g\mathbf{W}_\lambda$. Then g would disappear from all equations except for an overall factor g^{-2} in (4.9). Because this notation is rare in the literature, we do not use it.)

4.2 Feynman rules for the Yang–Mills field

Can one do Feynman perturbation theory with the Yang–Mills Lagrangian (4.5) and (4.9)? If so, what are the Feynman rules? The problem here is that $\mathbf{W}_\lambda$, like the electromagnetic potential $\mathscr{A}_\lambda$, transforms with an inhomogeneous term in (4.4), and so cannot be determined uniquely by the field equations. In order to quantize, one must 'fix a gauge', that is to say break the gauge-invariance. This might be done for example, in analogy with (3.22), by adding to the Lagrangian density a term

$$\mathscr{L}_G = -\tfrac{1}{2}\xi^{-1}(\partial\cdot\mathbf{W})^2. \tag{4.10}$$

In this case, however, in contrast to electrodynamics, it turns out that the addition of (4.10) by itself is *not* correct.

A simple example shows the error in using the Feynman rules extracted from (4.9) and (4.10) as they stand. The $\mathbf{W}_\lambda$ propagator is, as for photons, given by (3.24) (with an additional factor $\delta_{\alpha\beta}$, where $\alpha, \beta = 1, 2, 3$ are iso-vector indices). Return to fig. 3.1, for which we checked unitarity in the case of electrodynamics. But now let the zig-zag lines represent Yang–Mills quanta, and the solid line represent an iso-spinor. If we attempt to generalize the argument of §3.3, we will meet, in the analogue of (3.13), the commutator

$$\delta(x_0 - y_0)\,[J_0^\alpha(x), J_\nu^\beta(y)], \tag{4.11}$$

where in this chapter J_λ^α is the iso-spin current appropriate to (4.9). This commutator is *not* zero. In fact, the iso-spin operators

$$T^\alpha = \int \mathrm{d}^3\mathbf{x}\, J_0^\alpha(\mathbf{x}, t) \tag{4.12}$$

obey the commutation relations

$$[T^\alpha, T^\beta] = \mathrm{i}\epsilon^{\alpha\beta\gamma}T^\gamma. \tag{4.13}$$

Thus the argument breaks down at this crucial point and the Feynman rules taken from (4.9) and (4.10) are inconsistent with unitarity.

It turns out that the simple Feynman rules can be corrected by the addition of extra compensating diagrams. We postpone discussion of this rather technical point until chapter 11. The Lagrangian given by

(4.5), (4.9) and (4.10) appears to be a renormalizable one. It has no dimensional coupling constants, and the propagator (3.24) contains no $k_\lambda k_\nu / M^2$ term. We will see in chapter 11 that this appearance is not misleading: the extra compensating diagrams do not upset the renormalizability.

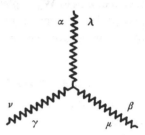

FIGURE 4.1. $3W$ vertex.

Although we do not yet have the complete Feynman rules, we can read off some of the rules from (4.5) and (4.9). For example, (4.9) implies that the three-vertex defined in fig. 4.1 contributes a factor (the momenta going into the vertex)

$$- i g \epsilon^{\alpha\beta\gamma} [(p-q)_\lambda \, g_{\mu\nu} + (q-k)_\mu \, g_{\nu\lambda} + (k-p)_\nu \, g_{\lambda\mu}]. \qquad (4.14)$$

Let us see how (4.14) can be interpreted as an interaction between a charged spin-1 particle and a photon. We temporarily suppose the former to have a mass M, which of course is not really present in the theory as it stands. Choose W_λ^3 to be the photon field, and define the charged fields

$$W_\lambda^\pm = 2^{-\frac{1}{2}} (W_\lambda^1 \pm i W_\lambda^2). \qquad (4.15)$$

Then the $\epsilon^{\alpha\beta\gamma}$ factor gives, for example,

$$\epsilon^{3-+} = \tfrac{1}{2} i (\epsilon^{312} - \epsilon^{321}) = i. \qquad (4.16)$$

In a suitable Lorentz frame, we set (with $|\mathbf{k}| \ll M$)

$$p \simeq \left(M + \frac{\mathbf{k}^2}{8M}, \ -\tfrac{1}{2}\mathbf{k} \right), \quad q \simeq \left(-M - \frac{\mathbf{k}^2}{8M}, \ -\tfrac{1}{2}\mathbf{k} \right), \quad k = (0, \mathbf{k}),$$

and the polarization vectors of the charged particles are

$$e = (-\mathbf{k} \cdot \mathbf{e}/2M, \mathbf{e}), \quad e' = (\mathbf{k} \cdot \mathbf{e}'/2M, \mathbf{e}'),$$

where $p \cdot e \simeq q \cdot e' \simeq 0$ as required. Then, contracting (4.14) with $e_\mu e'_\nu$ there results

$$- 2gM\mathbf{e} \cdot \mathbf{e}' + \tfrac{3}{2}gM^{-1}(\mathbf{k} \cdot \mathbf{e} \, \mathbf{k} \cdot \mathbf{e}' - \tfrac{1}{6}\mathbf{k}^2 \mathbf{e} \cdot \mathbf{e}') \quad (\lambda = 0) \qquad (4.17)$$

and

$$- 2g(\mathbf{k} \cdot \mathbf{e} \, \mathbf{e}' - \mathbf{k} \cdot \mathbf{e}' \mathbf{e}) \quad (\lambda \neq 0). \qquad (4.18)$$

Expression (4.17) represents an electric charge g and an electric quadrupole moment of strength $\tfrac{1}{4}gM^{-2}$; while (4.18) represents a magnetic moment of strength gM^{-1}. These particular values are selected by the symmetry of (4.14) and by its not containing powers of momenta higher than the first.

4.3　General Yang–Mills theories

It is not difficult to generalize the Yang–Mills formalism for a general compact semi-simple Lie group G of order N. In subsequent chapters we shall sometimes find it actually simpler to deal with the general case than with a special one.

The vector fields W_λ^α ($\alpha = 1, 2, \ldots, N$) belong to the regular representation of G (that is, the representation to which the generators belong). There may be fermion fields ψ belonging to any representation of G, in which the generators are represented by N Hermitian matrices t^α. These obey commutation relations

$$[t^\alpha, t^\beta] = \mathrm{i} f^{\alpha\beta\gamma} t^\gamma, \qquad (4.19)$$

in which the coefficients $f^{\alpha\beta\gamma}$ are the structure constants of G. We normalize the regular representation so that $f^{\alpha\beta\gamma}$ is totally antisymmetric (see Hamermesh 1960: 310; Gilmore 1974: 259). They obey the Jacobi identities

$$f^{\theta\alpha\delta} f^{\delta\beta\gamma} + f^{\theta\beta\delta} f^{\delta\gamma\alpha} + f^{\theta\gamma\delta} f^{\delta\alpha\beta} = 0. \qquad (4.20)$$

With these definitions, we simply have to write t^α for $\frac{1}{2}\tau^\alpha$ and $f^{\alpha\beta\gamma}$ for $\epsilon^{\alpha\beta\gamma}$ in the previous equations. For example, (4.4) becomes

$$W_\lambda^\alpha \to W_\lambda^\alpha + \partial_\lambda \omega^\alpha - g f^{\alpha\beta\gamma} \omega^\beta W_\lambda^\gamma, \qquad (4.21)$$

(4.3 a) becomes

$$D_\lambda = \partial_\lambda - \mathrm{i} g t^\alpha W_\lambda^\alpha, \qquad (4.22)$$

and (4.7) becomes

$$F_{\lambda\nu}^\alpha = \partial_\lambda W_\nu^\alpha - \partial_\nu W_\lambda^\alpha + g f^{\alpha\beta\gamma} W_\lambda^\beta W_\nu^\gamma. \qquad (4.23)$$

We have described the Yang–Mills field interacting with a fermion field as an example. It could equally well interact with a boson field, and it has its own self-interaction even in the absence of any other fields.

4.4　Infra-red difficulties

The Yang–Mills field cannot have any direct physical application, since gauge-invariance requires that none of the fields W_λ^α has mass terms, and only one massless spin-1 particle, the photon, is known in nature. What is more, it is not clear that the Yang–Mills theory could possibly have any direct interpretation in any case. The obstacle lies in infra-red divergences.

In quantum electrodynamics, certain integrals diverge at low momenta because of the absence of a photon mass (see, for example, Jauch and Rohrlich 1955: §16.1). This fact is explained when it is realized that the probability for emission of a charged particle by itself is zero. Such a particle is always accompanied by an indeterminate number of soft photons. This is not as complicated as it sounds since the coherent soft-photon states are handled classically to a good approximation.

The problem in Yang–Mills theory is, at best, much more complicated, since any soft Yang–Mills particle can itself emit others, and so whole tree-graphs of soft particles are involved. I am not aware of any detailed investigation of this problem. Weinberg (1965*b*) has argued that there is a real difficulty here. On the other hand, Lee and Nauenberg (1964) have a general method (involving averages over final *and* initial states) for obtaining finite answers in cases of this kind.

There is another argument that can be brought to bear for the Yang–Mills field on its own, not interacting with any massive particles. Without a mass to set a scale, one can try to use the renormalization group (see §18.5) to find out how the effective coupling constant behaves for small energies. This method fails to show that the coupling constant tends to zero (as it does in the quantum electrodynamics of massless electrons, for instance). Strictly this proves nothing, but it does cast further doubt on the theory.

It may be, therefore, that the Yang–Mills theory is not a self-consistent one; or it may be just that the spectrum of physical states cannot be inferred from the free (i.e. quadratic) part of the Lagrangian. This question is not relevant to most of the subsequent chapters, as they will be concerned with modifications of the Yang–Mills theory. We will return to it briefly in §18.5.

5

Spontaneous breaking of symmetries

5.1 Symmetries in infinite systems

We described the Yang–Mills field theory in the last chapter, because we wanted charged spin-1 particles to mediate weak interactions and because we were trying to mimic and generalize the gauge-invariance of electrodynamics. The Yang–Mills theory will certainly not do as it stands, since its spin-1 particles are all massless. How can one get massive particles while retaining (in some sense) the gauge-invariance? One way of obtaining unsymmetric solutions to a symmetric theory is that which is often called 'spontaneous breaking' of symmetries. (The phrase was coined by Baker & Glashow, 1962.)

Suppose a Lagrangian is invariant under a group G of transformations. If there is a unique state of minimum energy (ground state), it must be a singlet under G. Alternatively, there might be a degenerate set of minimum energy states, which transform under G as members of a multiplet. If one arbitrarily selects one of these as 'the' ground state of the system, the symmetry is said to be spontaneously broken.

The first alternative normally occurs in the quantum mechanics of systems with a finite number of degrees of freedom. The second alternative, however, is quite common in many-particle systems. A ferromagnet affords a simple example. The Hamiltonian has rotational invariance, but the ground state has the spins aligned in some arbitrary direction. What is more, any higher state, built from the ground state by a finite number of excitations, shares its anisotropy.

It is essential that the magnet be, in principle, infinite. Otherwise there would be a zero angular momentum state with less energy than any other. For the infinite system, however, the moment of inertia is infinite, and the zero angular momentum state is degenerate with states of non-vanishing angular momentum. Mathematically, the infinite size of the system prevents there being a unitary operator connecting the different aligned ground states – they are in different Hilbert spaces.

5.2   Goldstone's model field theory

Relativistic quantum field theory is concerned with an infinite number of degrees of freedom; so here also spontaneous symmetry-breaking is possible. The idea of exploiting this possibility was Nambu's (1960). A model field theory which shows the spontaneous breaking at work was invented by Goldstone (1961).

Let

$$\phi = 2^{-\frac{1}{2}}(\phi_1 + i\phi_2) \tag{5.1}$$

be a complex scalar field with Lagrangian density

$$\partial_\lambda \phi^\dagger \, \partial^\lambda \phi - V(\phi) \tag{5.2}$$

in which the potential has the form

$$V = \tfrac{1}{2}\lambda^2 |\phi|^4 - \tfrac{1}{2}\mu^2 |\phi|^2. \tag{5.3}$$

This Lagrangian is invariant under global phase transformations

$$\phi \to e^{i\omega}\phi \tag{5.4}$$

(ω constant).

The term in λ^2 is a self-interaction. It is usual to choose $\lambda^2 > 0$, since (5.3) suggests that the energy would have no minimum for $\lambda^2 < 0$. If, in (5.3), $\mu^2 < 0$, it would be an ordinary mass term and there would be nothing noteworthy about the model. Suppose instead we choose $\mu^2 > 0$. The function $V(\phi_1, \phi_2)$ then has the form shown in fig. 5.1 (where we have ignored the fact that $\phi(x)$ is a field). The surface has a shape like the bottom of a wine bottle, with a local maximum at $\phi = 0$ and a minimum along the circle

$$\phi_1^2 + \phi_2^2 = \mu^2/\lambda^2. \tag{5.5}$$

Fig. 5.1 suggests that the ground state (the vacuum) should be associated with values of ϕ near the circle (5.5); so that the vacuum-expectation-value of ϕ should not be zero but should satisfy

$$|\langle 0| \phi |0\rangle|^2 = \mu^2/(2\lambda^2). \tag{5.6}$$

The phase of $\langle 0| \phi |0\rangle$ is arbitrary, but we can choose the 1-, 2-axes so that

$$\langle 0| \phi_1 |0\rangle = \mu/\lambda, \quad \langle 0| \phi_2 |0\rangle = 0. \tag{5.7}$$

If this is true, the vacuum state $|0\rangle$ is clearly not invariant under the transformations (5.4).

Two points should be stressed about this reasoning. First, the argument for (5.6) from fig. 5.1 is only a suggestive one, since we have

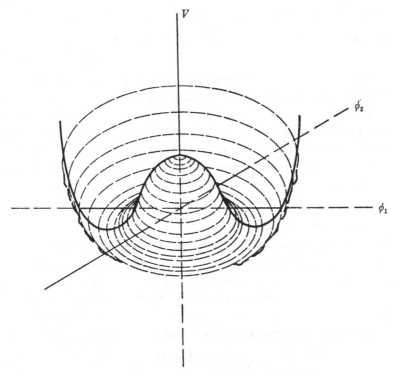

FIGURE 5.1. The potential surface $V(\phi_1, \phi_2)$.

totally ignored the fact that $\phi(x)$ is a field. Second, if ϕ_1 and ϕ_2 were not fields but were co-ordinates of a particle, the argument would be *wrong*: the ground state would be the symmetric one of zero angular momentum.

The correct attitude to (5.6) is that we try the *assumption* that

$$\langle 0| \phi(x) |0\rangle \neq 0, \tag{5.8}$$

and we verify that we meet no contradiction, at least within the restricted context of perturbation theory. A rigorous proof that (5.8) is true, or that it is not inconsistent, is beyond us at present. In § 14.5 we will reconsider the actual value assigned to (5.6). The stated value proves to be correct, by definition, provided that (5.2) is the renormalized Lagrangian.

All this being said, how do we calculate with (5.2) (for $\mu^2 > 0$)? It is simplest to define a new field,

$$\phi_1' = \phi_1 - \langle 0| \phi_1 |0\rangle, \tag{5.9}$$

which has zero vacuum-expectation-value. Then (5.2) is rewritten

$$\tfrac{1}{2}(\partial_\lambda \phi_1')^2 + \tfrac{1}{2}(\partial_\lambda \phi_2)^2 - \tfrac{1}{2}\mu^2\phi_1'^2 - \tfrac{1}{2}\mu\lambda\phi_1'(\phi_1'^2 + \phi_2^2)$$
$$- \tfrac{1}{8}\lambda^2(\phi_1'^2 + \phi_2^2)^2. \quad (5.10)$$

The hope now is that, in terms of ϕ_1', the Lagrangian (5.10) can be interpreted in the normal way, with no more concern about the properties of the vacuum. In this spirit we observe that ϕ_1' represents a particle with a real mass μ, but ϕ_2 is massless. The reason for this is intuitively clear from fig. 5.1: ϕ_1' corresponds to radial oscillations about a point on the minimum circle (5.5); ϕ_2 corresponds to the zero-frequency motion around this circle.

5.3   Finite volume effects

In §5.1 we emphasized the importance of having, in principle, an infinite number of degrees of freedom. One might wonder what would be the effect of a finite universe on the relativistic Goldstone model.

To estimate orders of magnitude let us take a finite volume V with periodic boundary conditions. The volume integral of the kinetic term in the Lagrangian density (5.2) can then be written as a Fourier sum

$$\sum_{\mathbf{k}} (\dot{\phi}_{\mathbf{k}}^\dagger \dot{\phi}_{\mathbf{k}} - \mathbf{k}^2\phi_{\mathbf{k}}^\dagger \phi_{\mathbf{k}}). \quad (5.11)$$

We are interested in the $\mathbf{k} = 0$ mode, given approximately by

$$\phi_0 = (\tfrac{1}{2}V)^{\frac{1}{2}}\mu\lambda^{-1}e^{i\theta}, \quad (5.12)$$

where θ is time-dependent. The contribution of this to (5.11) is

$$\tfrac{1}{2}V\mu^2\lambda^{-2}\dot{\theta}^2. \quad (5.13)$$

Setting $V\mu^2\lambda^{-2}\dot{\theta} \sim h$

we obtain a characteristic period of order

$$\hbar^{-1}V\mu^2\lambda^{-2}. \quad (5.14)$$

If we put $V \sim (c/H)^3,$

where H is Hubble's constant, (5.14) becomes

$$(\mu/\hbar H)^2 (\lambda/\hbar c)^{-2} H^{-1}, \quad (5.15)$$

where μ is expressed in energy units and $\lambda/\hbar c$ is dimensionless. It is clear that, for any sensible values of μ and λ, the period (5.15) is so much greater than H^{-1} as to be entirely irrelevant. The frequency of rotation from one 'vacuum' to another is negligibly slow.

5.4 Goldstone's theorem

In the model described in §5.2, an important role is played by ϕ_1 – a field with the quantum numbers of the vacuum whose vacuum-expectation-value is not zero. It seems to be necessary to have such a field in order to do simple perturbation-theory calculations in field theories with spontaneous symmetry-breaking. But there is no necessity in principle for such a field. Instead there might be, for example, a bilinear operator with non-zero vacuum-expectation-value.

Nevertheless, Goldstone's model illustrates one general property of spontaneous symmetry-breaking: the necessity for massless, spinless particles (or excitations). Inspired by Nambu's (1960) ideas, Goldstone first appreciated the importance of these bosons (Goldstone 1961, Goldstone, Salam and Weinberg 1962). We sketch a general proof of the theorem in a form appropriate to relativistic cases (Gilbert 1964). Let the symmetry group G of the Lagrangian be associated with a set of Noether currents $j_\lambda^\alpha(x)$ which are divergenceless,

$$\partial^\lambda j_\lambda^\alpha = 0. \tag{5.16}$$

As the symmetry is spontaneously broken, suppose that there is some operator O_1, which is not an invariant under G, but whose vacuum-expectation-value is non-zero. Since O_1 is not invariant, there must be something that transforms into it. Therefore there is another operator O_2 such that

$$\int d^3x\,[j_0^\alpha(\mathbf{x},t), O_2] = iO_1 \tag{5.17}$$

for at least one value of α.

Construct the function

$$F_\lambda(q) = \int d^4x\, e^{iq\cdot x}\langle 0|\,[j_\lambda^\alpha(x), O_2]|0\rangle. \tag{5.18}$$

O_2, like O_1, must be a Lorentz scalar, so F_λ has the form

$$F_\lambda(q) = q_\lambda F(q^2) + \theta(q_0)\,q_\lambda G(q^2). \tag{5.19}$$

Equation (5.16) gives

$$q^2 F(q^2) + \theta(q_0)\,q^2\,G(q^2) = 0; \tag{5.20}$$

and so $\qquad F(q^2) = A\delta(q^2), \quad G(q^2) = B\delta(q^2). \tag{5.21}$

Equation (5.17) implies that at least one of A and B is non-zero.

Now insert a complete set of states

$$\sum_n |n\rangle (2E_n)^{-1}\langle n| = 1 \tag{5.22}$$

between the operators in the commutator in (5.18). From (5.21) at least one of these states has a 4-momentum $p^{(n)}$ satisfying $(p^{(n)})^2 = 0$. This state is the massless 'Goldstone boson' whose existence was to be established. It is a spinless state because O_2 is a scalar.

Incidentally, this proof implies that $\langle 0|j_\lambda(x)|n\rangle$ is not zero. By Lorentz invariance, we can write

$$\langle 0|j_\lambda(0)|n\rangle = if p_\lambda^{(n)}, \tag{5.23}$$

where f is a non-zero constant. This equation is characteristic of spontaneous symmetry-breaking.

The proof of Goldstone's theorem for non-relativistic systems is more complicated. We refer the reader to the review by Guralnik, Hagen and Kibble (1968), and content ourselves with discussion of a simple example.

Goldstone excitations are easy to understand in a ferromagnet. Consider a spin-wave (magnon) of very long wave-length λ. Over regions of size small compared with λ, one has approximately a ground state with magnetization in a certain direction; but this direction varies slowly from region to region. It requires vanishingly little energy to excite this situation, provided the forces between spins are of finite range; and so the frequency of the spin-waves tends to zero with increasing wave-length. The proviso about finite-range forces is very important (see § 6.1).

Let us return briefly to the model of § 5.2. The current corresponding to the phase invariance is

$$j_\lambda = \phi_1 \partial_\lambda \phi_2 - \phi_2 \partial_\lambda \phi_1. \tag{5.24}$$

In terms of the field ϕ_1', this becomes

$$j_\lambda = \phi_1' \partial_\lambda \phi_2 - \phi_2 \partial_\lambda \phi_1' + (\mu/\lambda)\, \partial_\lambda \phi_2. \tag{5.25}$$

The last term in this equation helps us to understand (5.23), since ϕ_2 is the field which annihilates the Goldstone state $|n\rangle$.

We described the model for the simple case in which (ϕ_1, ϕ_2) was a two-dimensional representation of $U(1)$. It is easy to generalize to a group G, with generators Q^α ($\alpha = 1, ..., N$), of which ϕ is a real r-dimensional representation in which Q^α is represented by iv^α (each v^α being a real, antisymmetric $r \times r$ matrix). Let

$$\langle 0|\phi|0\rangle = F, \tag{5.26}$$

where F is a constant r-component vector in the representation space. The vectors

$$v^\alpha F \quad (\alpha = 1, ..., N) \tag{5.27}$$

span a subspace of dimension $n \leqslant r - 1$ (n cannot equal r because the subspace is orthogonal to F). The little-group G_F of F in G is unbroken so far as (5.26) goes.

The invariance under G of the potential $V(\phi)$ (generalizing (5.3)) tells us that

$$\frac{\partial V}{\partial \phi_i} v_{ij}^\alpha \phi_j = 0 \quad (\alpha = 1, ..., N). \tag{5.28}$$

We assume that

$$\left[\frac{\partial V}{\partial \phi}\right]_{\phi = F} = 0. \tag{5.29}$$

Then differentiating (5.28) gives

$$\left[\frac{\partial^2 V}{\partial \phi_k \partial \phi_i}\right]_{\phi = F} v_{ij}^\alpha F_j = 0. \tag{5.30}$$

The square bracket is the mass-matrix for the bosons, and (5.30) states that it has no elements in the subspace defined by (5.27). So there are n zero eigenvalues, and these correspond to n Goldstone bosons. The remaining $(r - n)$ particles have, in general, non-zero masses.

5.5   Soft pions

Although spontaneous breaking of symmetries occurs in non-relativistic many-particle systems, it seems to have no place in elementary particle physics since massless spin-0 bosons are not found in nature. Nevertheless, the notion of spontaneous breaking has proved useful in particle physics. The idea is that the non-strange weak axial-vector currents ($J_{5\lambda}^\alpha$, $\alpha = 1, 2, 3$, in the notation of §2.2) are *nearly* divergenceless and the pions (by far the lightest hadrons) are *nearly* Goldstone bosons. This hypothesis, it is argued, accounts for the success of the Goldberger–Treiman relation and other predictions of $PCAC$ (partially conserved axial current). The theory is outside the scope of this book. The reader is referred to Adler and Dashen (1968), chapter 2. We shall, however, indicate how Goldstone's model is modified to give the Goldstone bosons a small mass.

Supplement the potential V in (5.3) with a small extra term

$$-\epsilon \phi_1, \tag{5.31}$$

so that the phase invariance under (5.4) is broken. The potential surface now has the form indicated in figure 5.2 instead of fig. 5.1. There is a *unique* minimum at

$$\phi_1 = f, \quad \phi_2 = 0, \tag{5.32}$$

where

$$\lambda^2 f^3 - \mu^2 f - 2\epsilon = 0. \tag{5.33}$$

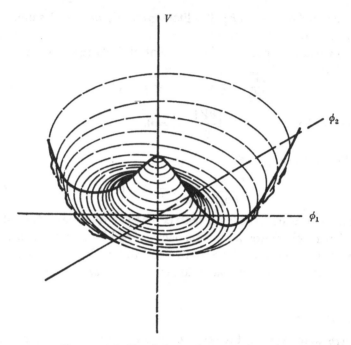

FIGURE 5.2. The deformed potential surface.

Proceeding as in §5.2, but using (5.33) instead of (5.7), one obtains in (5.10) an extra term

$$-\tfrac{1}{4}(\lambda^2 f^2 - \mu^2)\,\phi_2^2 = -(\epsilon/2f)\,\phi_2^2 \simeq -(\epsilon\lambda/2\mu)\,\phi_2^2. \qquad (5.34)$$

This is the small mass-term for the Goldstone field.

Under an infinitesimal phase transformation (5.4), the Lagrangian changes, because of (5.31), by

$$\epsilon\delta\phi_1 = -\epsilon\omega\phi_2. \qquad (5.35)$$

It follows that
$$\partial\cdot j = -\epsilon\phi_2, \qquad (5.36)$$

as can also be deduced from (5.24) and the equations of motion. Thus (5.34) relates the mass to the departure from divergencelessness.

To extend this formalism to the physical situation of $PCAC$ (Gell-Mann and Lévy 1960), we choose for the original invariance group $SU(2)_{\mathrm{L}} \times SU(2)_{\mathrm{R}}$, associated with the currents L_λ^α, R_λ^α ($\alpha = 1, 2, 3$) defined in §2.2. For the boson fields we choose a real 4-dimensional representation (assuming the pions to have fundamental fields for the purposes of the model)

$$(\sigma, \pi_1, \pi_2, \pi_3) = (\sigma, \boldsymbol{\pi}). \qquad (5.37)$$

Since $SU(2) \times SU(2)$ is locally isomorphic to $O(4)$, we may think of (5.37) as a 4-vector under $O(4)$. Under the iso-spin subgroup, $\boldsymbol{\pi}$ transforms as an iso-vector, σ as an iso-scalar. The action of the remaining generators is illustrated by the commutators

$$\left.\begin{aligned} [J_{50}^{\alpha}(\mathbf{x},t), \sigma(\mathbf{y},t)] &= i\pi^{\alpha}(\mathbf{x},t)\,\delta^3(\mathbf{x}-\mathbf{y}), \\ [J_{50}^{\alpha}(\mathbf{x},t), \pi^{\beta}(\mathbf{y},t)] &= -i\delta^{\alpha\beta}\sigma(\mathbf{x},t)\,\delta^3(\mathbf{x}-\mathbf{y}). \end{aligned}\right\} \qquad (5.38)$$

σ and $\boldsymbol{\pi}$ clearly have opposite parity, and we may choose σ to be a scalar. The symmetry-breaking term, like (5.31), is taken to be

$$-\epsilon\sigma, \qquad (5.39)$$

since this conserves both parity and iso-spin. The potential V is similar to fig. 5.2 (but generalized to 4 dimensions). It has a unique minimum at

$$(\sigma, \boldsymbol{\pi}) = (f_\pi, \mathbf{0}). \qquad (5.40)$$

The analogues of (5.34) and (5.36) are

$$-(\epsilon/2f_\pi)\,\boldsymbol{\pi}^2 \qquad (5.41)$$

and

$$\partial \cdot J_5^{\alpha} = -\epsilon\pi^{\alpha}, \qquad (5.42)$$

the latter being the operator form of $PCAC$. The results of soft pion theory are obtained from (5.41) and (5.42), assuming ϵ to be small. From the equation

$$\langle 0| J_{5\lambda}^{\alpha}|\pi^{\beta}\rangle = i\delta^{\alpha\beta}f_\pi p_\lambda \qquad (5.43)$$

(like (5.23)) and the observed pion decay rate, one obtains

$$|f_\pi| \simeq 90\,\text{MeV}. \qquad (5.44)$$

The above brief mention of $PCAC$ is intended to suggest that spontaneous symmetry-breaking may have applications in particle physics. It is *not* directly relevant to the subject of local gauge theories, since in these one cannot allow an explicit symmetry-breaking term like (5.31) or (5.39), however small.

6
Spontaneous breaking of local gauge symmetries

6.1 Non-relativistic examples

In chapter 4, we said that non-abelian local gauge theories have no simple application in physics since they require too many massless spin-1 particles (or, if not, are not susceptible to conventional perturbation theory). In chapter 5, we argued that spontaneous breaking of a global symmetry entails massless spin-0 quanta, which also are not observed in particle physics. The remarkable thing is that the spontaneous breaking of *local* gauge symmetry leads to a situation in which both these difficulties can be simultaneously abolished. The spin-1 particles can acquire mass, and the Goldstone fields turn out not to represent spin-0 particles.

We mention first how this happens in non-relativistic many-body theory. In §5.1 we gave a physical argument why the frequency of spin-waves in a ferromagnet goes to zero with the wave-number. This argument depended upon the absence of long-range forces between the spins. But the example that first attracted the attention of particle physicists (Nambu 1960, Nambu and Jona-Lasinio 1961) was not ferromagnetism but superconductivity. In BCS theory (Bardeen, Cooper and Schrieffer 1957, Anderson 1958) the superconducting state is reached by the spontaneous breaking of the phase invariance which is associated with conservation of electron-number (in the pseudo-spin formulation of Anderson 1959, the broken symmetry group is $SU(2)$). In this case, however, there is a long-range Coulomb force present (though Nambu, 1960, did not copy this feature in his relativistic model).

Superconductivity is complicated, partly, by the fact that electrons are fermions. A simpler system, with some features in common is a degenerate Bose gas with a potential whose Fourier transform is $V(\mathbf{k})$. In the ground state, the condensate field ϕ satisfies

$$\langle 0| \phi |0 \rangle = F, \tag{6.1}$$

where F is a complex number, and $|F|^2$ is the expectation-value of the particle number density n. The phonon frequency ω at wave-number $\mathbf{k}$ is given by

$$\omega^2 = \frac{\mathbf{k}^2}{2m}\left[\frac{\mathbf{k}^2}{2m} + 2|F|^2 V(\mathbf{k})\right]. \tag{6.2}$$

If V is short-range, in the sense that

$$\mathbf{k}^2 V(\mathbf{k}) \to 0 \quad \text{as} \quad \mathbf{k} \to 0, \tag{6.3}$$

then the phonons are true Goldstone excitations with

$$\omega \to 0 \quad \text{as} \quad \mathbf{k} \to 0. \tag{6.4}$$

If on the other hand, V is a (repulsive) Coulomb potential $e^2/\mathbf{k}^2$, then

$$\omega \to \omega_{\mathrm{P}} \quad \text{as} \quad \mathbf{k} \to 0, \tag{6.5}$$

where ω_{P} is the plasma frequency given by

$$\omega_{\mathrm{P}}^2 = |F|^2 e^2/m = \langle n \rangle e^2/m. \tag{6.6}$$

In this case, Goldstone's theorem is evaded. (For further details, see the review of Guralnik, Hagen and Kibble 1968.)

Superconductivity can be covered by this model if ϕ is identified with the Cooper-pair condensate wave-function, as in the Ginzburg–Landau theory (see, for example, Rose-Innes and Rhoderick 1969).

6.2 Higgs' model

(For a careful review of the principles involved in this section, see Bernstein 1974.)

Anderson (1963) pointed out that, in the presence of the long-range force of electromagnetism, Goldstone's theorem might be as inapplicable in a relativistic as in a non-relativistic context (see also Lange 1966). Higgs (1964a, b, 1966) invented a model which shows this very simply (see also Englert and Brout 1964, Guralnik, Hagen and Kibble 1964). It turns out to be simpler to focus attention, not so much on the long-range character, as on the local gauge-invariance of electromagnetism (of course these two are intimately related). We now describe Higgs' simple abelian model. We show later this model does not provide anything practically useful, but it does teach us how to construct useful non-abelian models.

Generalize the gauge transformation (5.4) to the local form

$$\phi \to \phi \exp[ie\omega(x)], \tag{6.7}$$

like (3.21). (The coupling constant e is inserted in (6.7) to conform with common usage.) Then the Lagrangian density (5.2) must be replaced by

$$|(\partial_\lambda - ie\mathscr{A}_\lambda)\phi|^2 - V(\phi) - \tfrac{1}{2}(\partial_\lambda\mathscr{A}_\nu - \partial_\nu\mathscr{A}_\lambda)\partial^\lambda\mathscr{A}^\nu, \qquad (6.8)$$

like (3.20).

Proceeding just as in §5.2, we assume that

$$\langle 0|\phi_1|0\rangle = f \neq 0 \qquad (6.9)$$

and define

$$\phi_1 = f + \phi_1'. \qquad (6.10)$$

Transcribing (6.8) in terms of ϕ_1', the following two terms (among others) appear

$$\tfrac{1}{2}e^2 f^2 \mathscr{A}_\lambda \mathscr{A}^\lambda, \qquad (6.11)$$

$$-ef\mathscr{A}_\lambda \partial^\lambda \phi_2. \qquad (6.12)$$

The obvious interpretation of (6.11), which turns out to be correct, is that the 'photon' has acquired a mass

$$M = ef. \qquad (6.13)$$

(Since $\mathscr{A}_\lambda \mathscr{A}^\lambda = \mathscr{A}_0^2 - \mathscr{A}_i^2$, the sign of (6.11) is correct for the space-components of $\mathscr{A}_\lambda$.) A similarity to (6.6) is evident.

The interpretation of (6.12) is less clear. It mixes ϕ_2 with $\mathscr{A}_\lambda$, and so obscures the physical interpretation of each of these fields. To elucidate the role of ϕ_2, we express the gauge transformation (6.7) in terms of ϕ_1' and ϕ_2. For infinitesimal ω it becomes

$$\left.\begin{aligned} \phi_1' &\to \phi_1' - e\omega\phi_2, \\ \phi_2 &\to \phi_2 + e\omega\phi_1' + e\omega f. \end{aligned}\right\} \qquad (6.14)$$

Thus ϕ_2, like $\mathscr{A}_\lambda$, undergoes an *inhomogeneous* transformation. Neither ϕ_2 nor $\mathscr{A}_\lambda$, therefore, can individually have a direct physical meaning.

The simplest procedure, apparently, makes use of the gauge freedom, to arrange that

$$\phi_2 = 0. \qquad (6.15)$$

This condition fixes the gauge (rather as (3.7) fixed the gauge of e_λ). The mixing term (6.12) then disappears. The Goldstone boson (which we expected ϕ_2 to describe) has proved to be non-existent. The physical interpretation is now clear. There is an ordinary massive vector meson $\mathscr{A}_\lambda$, and a massive neutral scalar field ϕ_1' (with a mass $\mu = \lambda f$ as in (5.10)). They have the particular interaction terms

$$\tfrac{1}{2}e^2\phi_1'^2 \mathscr{A}_\lambda \mathscr{A}^\lambda + eM\phi_1' \mathscr{A}_\lambda \mathscr{A}^\lambda, \qquad (6.16)$$

but that, on the face of it, is all that is noteworthy about them. In particular, the propagator for $\mathscr{A}_\lambda$, coming from (6.8) and (6.11), is just the conventional propagator (2.26).

Two massless particles have been disposed of: the 'photon' has become massive, and the Goldstone boson has gone away. The total count of free-particle states has not changed. The original Lagrangian (6.8) seemed to contain two spin-0 particles and a massless photon with two spin-1 states – a total of four states. After the symmetry-breaking, we find a single spin-0 particle (ϕ_1') and a massive spin-1 meson with three spin-states. The total remains four. Somehow the Goldstone boson has been exchanged for the longitudinal polarization state of the vector meson.

6.3 't Hooft's gauges

The model described in §6.2 is superficially not a useful one, since, as explained in §2.3, the propagator (2.26) usually leads to non-renormalizable behaviour.

So, was the renormalizability of the initial Lagrangian (6.8) spoilt by the spontaneous symmetry-breaking? Weinberg (1967) and Salam (1968) conjectured that it was not, and 't Hooft (1971a) proved that it was not. 't Hooft's crucial discovery was that it was useful to adopt a more general gauge than (6.15), in spite of the latter's simplicity. Veltman has said that the advantage of so doing is a little like the advantage of going from real to complex variables in function theory.

't Hooft's gauges are defined by adding a term to the Lagrangian which generalizes (3.22) or (4.10). For the model of §6.2, the gauge-fixing term is
$$\mathscr{L}_G = -\tfrac{1}{2}\xi^{-1}(\partial\cdot\mathscr{A} + \xi M\phi_2)^2, \tag{6.17}$$
where ξ is a parameter which determines the particular gauge. $\mathscr{L}_G$ is chosen so as, for convenience, to cancel the awkward mixing term (6.12). The vector-meson propagator, coming from (6.8), (6.11) and (6.17) is the inverse of
$$-(k^2 g_{\lambda\nu} - k_\lambda k_\nu) + M^2 g_{\lambda\nu} - \xi^{-1}k_\lambda k_\nu. \tag{6.18}$$
Writing (6.18) in terms of projection operators
$$-(k^2 - M^2)\left(g_{\lambda\nu} - \frac{k_\lambda k_\nu}{k^2}\right) + (M^2 - \xi^{-1}k^2)\frac{k_\lambda k_\nu}{k^2},$$
the inverse is seen to be
$$-(k^2 - M^2)^{-1}\left(g_{\lambda\nu} - \frac{k_\lambda k_\nu}{k^2}\right) + (M^2 - \xi^{-1}k^2)^{-1}\frac{k_\lambda k_\nu}{k^2}$$

or $\qquad [-g_{\lambda\nu} + (1-\xi)\,k_\lambda k_\nu (k^2 - \xi M^2)^{-1}]\,(k^2 - M^2 + i\epsilon)^{-1}.$ $\qquad$ (6.19)

The field ϕ_2 gets a mass from (6.17), and its propagator is

$$(k^2 - \xi M^2)^{-1}. \qquad (6.20)$$

By simple power counting, the propagator (6.19) is expected to give a renormalizable theory for any finite value of ξ. The fields $\mathscr{A}_\lambda$ and ϕ_2 separately have no physical significance. In particular, the poles at $k^2 = \xi M^2$ in (6.19) and (6.20) are unphysical, and are cancelled out in any S-matrix element (which is why they are not defined by an $i\epsilon$ or other prescription).

Some special values of ξ are noteworthy. $\xi = 1$ gives the Feynman gauge, like that often used in electrodynamics. $\xi = 0$ gives the Landau gauge, in which the numerator of (6.19) is the transverse projection operator. As $\xi \to \infty$, (6.19) approaches the ordinary, apparently non-renormalizable, propagator (2.26); and the ϕ_2 propagator (6.20) tends to zero, suggesting that this field effectively drops out of the Feynman integrals.

The situation, then, is this. Gauges with ξ finite are manifestly renormalizable, but they are complicated in that five fields, $\mathscr{A}_\lambda$ and ϕ_2, are used to describe the three spin-states of the vector particle. The gauge $\xi \to \infty$, on the other hand, is simple to interpret physically, but it is not obviously renormalizable. This gauge is often called the unitary gauge. Gauge-invariance is expected to guarantee that S-matrix elements (but *not* Green's functions in general) are independent of ξ (see chapter 11), so one can have the advantage of both sorts of gauge. In practice, it is probably more convenient to use a renormalizable gauge (say $\xi = 1$ or 0) where all the usual methods of renormalization theory are applicable. Because S-matrix elements are independent of ξ, they will be physically sensible and unitary. *Exact gauge-invariance is vital to assure independence of ξ.*

To describe the polarization state of an incoming or outgoing spin-1 particle, we must assign a wave-function for the five fields $\mathscr{A}_\lambda$, ϕ_2. A simple choice is

$$(e_\lambda; 0), \qquad (6.21)$$

where e_λ is an ordinary transverse polarization vector. The gauge-invariance under (3.6) and (6.14) allows us to replace (6.21) by

$$(e_\lambda + \omega k_\lambda; M\omega), \qquad (6.22)$$

where ω is an arbitrary number. Expressions (6.21) and (6.22) describe the same physical state.

6.4   Neutral vector fields

At this point we have to admit that the model described in §6.2 is after all of theoretical interest only. The reason is that it has been known for some time that a massive neutral vector meson interacting with a conserved current is renormalizable (Matthews 1949, Kamefuchi 1960, Salam 1960). This is because the dangerous term $k_\lambda k_\nu / M^2$ in the propagator (2.26) can be shown not to contribute to any S-matrix element. This fact could be verified by an adaptation of the argument for the removal of the t-dependent terms in (3.9). We will give an indirect proof, using the Higgs model of §6.2. This will help to point the contrast with the non-abelian case, to be discussed in the next section.

For the indirect proof, we supplement the Lagrangian (6.8) by adding, say, a charged fermion

$$i\overline{\psi}\gamma_\lambda(\partial^\lambda - ie'\mathscr{A}^\lambda)\,\psi - m\overline{\psi}\psi. \tag{6.23}$$

We have deliberately made the charge, e', here *different* from the charge e, of the ϕ field in (6.8). In abelian local gauge theories, there is nothing to relate the charges of different fields (see §7.2). The point of this rather artificial manoeuvre is that we can now let $e \to 0$ while keeping M and e' fixed. The field ϕ_1' in §6.2 then becomes a free field, and it can be disregarded. But the coupling between the (still massive) vector field and ψ remains, and the current is conserved. We are left with a massive vector particle coupled to ψ in a renormalizable way.

6.5   The general non-abelian case

Suppose we generalize the work of this chapter by substituting a non-abelian gauge transformation for the phase transformation of (6.7) (how to do this will be described in a moment). The strength of the couplings are then fixed once and for all by the factor g that occurs in $F^\alpha_{\lambda\nu}$ in (4.7). All multiplets with which W^α_λ interacts must have the same charge g. The device we used in §6.4 – letting the charge of the Higgs field tend to zero by itself – is impossible. In the non-abelian case, Higgs' mechanism generates a renormalizable theory, and it is of a new, nontrivial kind with one or more Higgs particles (like ϕ_1') playing an essential role.

We now show how to generalize Higgs' model for a non-abelian symmetry group (Kibble 1967). We go straight to the general case. This involves a certain amount of formalism, but it makes the principles clearer than would a particular example. We shall require the general formalism in chapter 15; but some readers may prefer to turn now to chapter 7, for which the contents of this section are not essential.

We combine the general Goldstone model of the end of §5.4 with the general Yang–Mills field of §4.3, so as to have local gauge-invariance under the group G.

We generalize the formalism of §4.3 in one respect: G may be a direct product of a number of subgroups. In this case the coupling constants for the different factor groups will in general be different. We use the notation g^α $(\alpha = 1, ..., N)$ for the coupling constants, where the g^α are equal for α running over the generators of any one factor group. This throws out the usual summation convention for repeated indices. Instead we make the convention that, where g^α occurs, a sum over α is understood only if there are *three* symbols bearing the affix α.

The complete Lagrangian density is

$$\tfrac{1}{2}[(\partial_\lambda + g^\alpha v^\alpha W^\alpha_\lambda)\phi]^2 - V(\phi) - \tfrac{1}{4}(F^\alpha_{\lambda\nu})^2, \qquad (6.24)$$

where $F^\alpha_{\lambda\nu}$ is defined in (4.23). Equation (6.24) is invariant under the local transformations (given in their infinitesimal form)

$$\phi \to \phi - g^\alpha v^\alpha \omega^\alpha \phi, \qquad (6.25)$$

$$W^\alpha_\lambda \to W^\alpha_\lambda + \partial_\lambda \omega^\alpha - g^\alpha f^{\alpha\beta\gamma} \omega^\beta W^\gamma_\lambda. \qquad (6.26)$$

The real, antisymmetric matrices v^α obey

$$[v^\alpha, v^\beta] = f^{\alpha\beta\gamma} v^\gamma. \qquad (6.27)$$

$V(\phi)$ is an invariant function containing quadratic, quartic and, possibly, cubic terms.

As in chapter 5, we assume that

$$\langle 0|\phi|0\rangle \equiv F \neq 0, \qquad (6.28)$$

and define $\qquad\qquad \phi = F + \phi'. \qquad (6.29)$

Expression (6.24) then provides a W mass term

$$\tfrac{1}{2}g^\alpha g^\beta(\tilde{F}\tilde{v}^\alpha v^\beta F)\, W^\alpha_\lambda W^{\lambda\beta} \qquad (6.30)$$

($\sim$ denotes the transpose of a matrix), and a mixing term

$$-g^\alpha \tilde{F}v^\alpha(\partial^\lambda \phi')\, W^\alpha_\lambda. \qquad (6.31)$$

't Hooft's gauge-fixing term generalizing (6.17) is

$$-\tfrac{1}{2}\xi^{-1}\sum_\alpha (\partial \cdot W^\alpha + \xi g^\alpha \tilde{F}v^\alpha \phi')^2, \qquad (6.32)$$

which cancels (6.31) and gives a 'mass' term

$$-\tfrac{1}{2}\xi \sum_\alpha (g^\alpha \tilde{F} v^\alpha \phi')^2. \qquad (6.33)$$

To describe in more detail how the group G is broken, consider the $N \times r$ matrix
$$g^\alpha (v^\alpha F)_i \quad (\alpha = 1, ..., N; i = 1, ..., r). \qquad (6.34)$$
Suppose it to have rank n, where $n \leqslant r-1$, $n \leqslant N$. There exist orthogonal matrices $E^{\alpha\beta}$ and C_{ij} such that

$$E^{\alpha\beta} g^\beta (v^\beta F)_i C_{ij} = M^\alpha \delta^\alpha_j, \qquad (6.35)$$

where the M^α are real numbers satisfying

$$M^\alpha = 0 \quad \text{for} \quad \alpha = n+1, ..., N. \qquad (6.36)$$

In order to find E and C, one may choose E to diagonalize the symmetric $N \times N$ matrix
$$g^\alpha (\tilde{F} \tilde{v}^\alpha v^\beta F) g^\beta \qquad (6.37)$$
and C to diagonalize the symmetric $r \times r$ matrix

$$\sum_\alpha (g^\alpha)^2 (v^\alpha F)_i (v^\alpha F)_j. \qquad (6.38)$$

The quantities $(M^\alpha)^2$ are the non-zero eigenvalues of each of these two square matrices. (To prove (6.35), note that the vectors $E^{\alpha\beta} g^\beta (v^\beta F)$, $\alpha = 1, ..., n$, form an orthogonal set.)

Now make a change of basis in the space of the regular representation of G, defining
$$\hat{W}^\alpha_\lambda = E^{\alpha\beta} W^\beta_\lambda. \qquad (6.39)$$

Similarly, make a change of basis in the space of the representation to which ϕ belongs, defining
$$\hat{\phi}_j = \phi_i C_{ij}. \qquad (6.40)$$

(These transformations may not be convenient in practice, but they help to exhibit the structure of the theory.)

In this notation (6.35) and (6.36) give

$$\hat{v}^\alpha \hat{F} = 0 \quad \text{for} \quad \alpha = n+1, ..., N. \qquad (6.41)$$

It follows that there is a subgroup of G given by the generators $\hat{Q}^\alpha$ ($\alpha = n+1, ..., N$) which is unbroken by (6.28). This is the little-group G_F. (It may be that $n = N$, so that there is no little-group.)

In the new basis, the mass term (6.30), by virtue of (6.35), becomes

$$\tfrac{1}{2} \sum_{\alpha=1}^{n} (M^\alpha)^2 \hat{W}^\alpha_\lambda \hat{W}^{\lambda\alpha} \qquad (6.42)$$

showing that the fields (6.39) are mass eigenstates.

The fields $\hat{\phi}_i\,(i = 1, ..., n)$ lie in the subspace of their representation space that is accessible from $\hat{F}$ by the action of the generators of G. These fields would have corresponded to Goldstone particles if the gauge symmetry had not been a local one. We call them 'Goldstone fields'. As it is, these fields transform inhomogeneously under (6.25) (using (6.29)); and so they have no physical significance separately from the $\hat{W}_\lambda^\alpha\,(\alpha = 1, ..., n)$. From (6.33), the Goldstone fields have a 'mass' term

$$-\tfrac{1}{2}\xi \sum_{i=1}^{n} M_i^2\, \hat{\phi}_i^2. \tag{6.43}$$

Finally, there are the fields $\hat{\phi}_i'\,(i = n+1, ..., r)$, which transform homogeneously under (6.25). They represent genuine spinless particles, which we call 'Higgs particles'. Equation (5.30) becomes

$$\left[\frac{\partial^2 V}{\partial \hat{\phi}_i\, \partial \hat{\phi}_j}\right]_{\phi=F} (\partial^\alpha \hat{F})_i = 0 \tag{6.44}$$

or, from (6.35),

$$\left[\frac{\partial^2 V}{\partial \hat{\phi}_i\, \partial \hat{\phi}_j}\right]_{\phi=F} = 0 \quad \text{for} \quad i,j = 1, ..., n. \tag{6.45}$$

Thus the second-order terms in the expansion of V are

$$\frac{1}{2} \sum_{i,j=n+1}^{r} \left[\frac{\partial^2 V}{\partial \hat{\phi}_i\, \partial \hat{\phi}_j}\right]_{\phi=F} \hat{\phi}_i'\, \hat{\phi}_j', \tag{6.46}$$

providing in general non-zero masses for the Higgs particles. This mass-matrix is positive provided that $\phi = F$ is a true minimum of V. The mass terms (6.46) may, if required, be diagonalized by an orthogonal transformation on $\hat{\phi}_i\,(i = n+1, ..., r)$ which does not interfere with (6.35).

The abelian model of §6.2 had $N = 1, r = 2, n = 1$, and there was no little-group G_F. In chapter 8, we shall encounter a case in which $N = 4, r = 4, n = 3$, and there is a one-parameter little-group $U(1)$.

For a general discussion of the group theory of symmetry-breaking, see Li (1974).

6.6   Current matrix-elements in Higgs theories

If a theory undergoes spontaneous symmetry-breaking, the current of the broken symmetry connects vacuum and one-particle states in a characteristic way. For simplicity, take a theory with a single conserved current j_λ:

$$\partial^\lambda j_\lambda = 0. \tag{6.47}$$

In Goldstone symmetry-breaking, j_λ connects the vacuum with the Goldstone-boson state, as in (5.23). In Higgs symmetry-breaking, j_λ connects the vacuum with a vector-meson state (Weinstein 1973b).

To see this, take the current corresponding to the Lagrangian (6.8):

$$j_\lambda = i(\phi^\dagger \partial_\lambda \phi - \phi \partial_\lambda \phi^\dagger) + 2e\phi^\dagger \phi \mathscr{A}_\lambda. \qquad (6.48)$$

The substitution (6.10) gives

$$j_\lambda = ef^2 \mathscr{A}_\lambda - f \partial_\lambda \phi_2 + \text{(non-linear terms)}. \qquad (6.49)$$

Equation (6.47) implies

$$0 = ef^2 \partial \cdot \mathscr{A} - f \,\square\, \phi_2 + \text{(non-linear terms)}, \qquad (6.50)$$

which allows ϕ_2 to be eliminated from (6.49). We then obtain, from (6.49) and (6.50),

$$\langle 0 | j_\lambda(0) | e, k \rangle = e^{-1}(g_{\lambda\nu} M^2 - k_\lambda k_\nu) e^\nu + \text{(terms of order } e^0), \quad (6.51)$$

where $|e, k\rangle$ is a spin-1 particle state with polarization vector e. This is the equation that replaces (5.23) and characterizes Higgs' mechanism.

The vacuum-expectation-value

$$(2\pi)^4 \int d^4x\, e^{-iq \cdot x} \langle 0 | T(j_\lambda(x) j_\nu(0)) | 0 \rangle \qquad (6.52)$$

acquires from (6.51) a pole contribution

$$i(M^2/e^2)(-M^2 g_{\lambda\nu} + q_\lambda q_\nu)(q^2 - M^2 + i\epsilon)^{-1}. \qquad (6.53)$$

In the limit $M \to 0$, this yields the pole contribution from the Goldstone matrix-element (5.23):

$$if^2 q_\lambda q_\nu (q^2 + i\epsilon)^{-1}. \qquad (6.54)$$

(There are non-pole additions to (6.53) and (6.54) which make them consistent with the conservation condition (6.47) applied to (6.52).)

6.7 High-energy behaviour of tree-graphs

Heavy vector mesons generally cause the Born approximation (and other tree-graphs) to contradict unitarity at high energies (§ 2.3). How is this avoided in gauge theories? In such a theory, the polarization state of a spin-1 vector can be represented by either of the 5-component quantities (6.21) or (6.22). If the former is used, the longitudinal polarization state requires the 4-vector (2.36), having the dangerous M^{-1} terms which generally signal bad high-energy behaviour. If (6.22) is employed, however, it can, by a suitable choice of ω, be brought to the form

$$(e'_\lambda; -1), \tag{6.55}$$

where $\qquad e'_0 = -M(k_0 + |\mathbf{k}|)^{-1}, \quad \mathbf{e}' = M\hat{\mathbf{k}}(k_0 + |\mathbf{k}|)^{-1}. \tag{6.56}$

The bad high-energy behaviour (equivalently, the singularity at $M = 0$) is absent here; so it must in fact cancel out in the (6.21) gauge also. The argument here exactly parallels the argument for renormalizability based on the propagator (6.19). Expression (6.21) corresponds to the unitary gauge $\xi \to \infty$.

The above argument is given in more detail by Bell (1973) and Schechter and Ueda (1973). A converse theorem has been proved: the only vector-meson theories with tree-graphs with good high-energy behaviour are Higgs theories (Cornwall, Levin and Tiktopoulos 1974; Llewellyn Smith 1973). If the underlying symmetry group G contains an abelian subgroup, the relevant Higgs field (if it transforms under that subgroup only) may be decoupled by the trick used in §6.4.

Another interesting point relates to the zero-mass limit of Higgs theories. If the symmetry-breaking parameter F in (6.28) is allowed to tend to zero the masses of the Yang–Mills particles become zero, but the Goldstone fields must also be interpreted as physical massless (spin-0) particles in the limit (the terms mixing them with the Yang–Mills field vanish in the limit). The nature of the limit can be seen from (6.55) and (6.56): as $M \to 0$ the longitudinal polarization mode becomes a pure scalar mode.

The pure Yang–Mills theory, with no massless spin-0 particles, is *not* obtainable as a zero-mass limit in any way.

The crucial point about the Higgs theories is that they *do have a smooth limit* as $M \to 0$. (Remember the singular M^{-1} terms, mentioned in §2.3, for vector mesons which were not part of a Higgs theory.)

6.8 Dynamical symmetry-breaking

In BCS theory, the energy-gap is calculated in terms of the interactions between electrons, whereas the vacuum-expectation-value of a Higgs field is an arbitrary constant independent of the properties of other particles. But does the Higgs field have to be a fundamental one, or could it be a phenomenological device for representing spin-0 states, constructed out of fermion fields for instance? Can a Higgs particle be a bound state? Certainly in BCS theory there is no fundamental scalar field (in §6.1, ϕ corresponded to Cooper pairs). There was no fundamental Goldstone field in the relativistic model of Nambu and Jona-Lasinio (1961).

Schwinger (1962a) has suggested that, in the photon self-energy

$$\Pi_{\lambda\nu}(k) = (g_{\lambda\nu}k^2 - k_\lambda k_\nu)\,\Pi(k^2),$$

the exact function $\Pi(k^2)$ might have a pole at $k^2 = 0$. Then the transverse part of the photon propagator,

$$\left(-g_{\lambda\nu} + \frac{k_\lambda k_\nu}{k^2}\right)[k^2 - k^2\,\Pi(k^2)]^{-1} \tag{6.57}$$

would have its pole shifted to the non-zero mass

$$\lim_{k^2 \to 0}[k^2\Pi(k^2)]. \tag{6.58}$$

This phenomenon does occur for quantum electrodynamics in spacetime of dimension 2 (Schwinger 1962b).

Englert and Brout (1964, 1974) and Jackiw and Johnson (1973) have re-examined Schwinger's idea, and indicated how the mechanism could work in 4 dimensions. With a single fermion field ψ, the effective Higgs operator must be $\bar\psi\psi$ or $\bar\psi\gamma_5\psi$. Since both of these are even under charge-conjugation, the current concerned must be an axial-vector (which is also even).

The model of Nambu and Jona-Lasinio (1961) was not renormalizable, and a cut-off was used in it. To make a renormalizable model, Cornwall and Norton (1973) introduce a second vector field to mediate the interaction between the fermions. They sum a series of graphs like that in fig. 6.1 (a). Alternatively, Jacobs (1974) has studied a direct four-fermion interaction in 2 dimensions (where it is renormalizable). The graphs summed are those in fig. 6.1 (b). This approximation is justified for an $SU(N)$ symmetry if N is large.

The idea of dynamical symmetry-breaking is very attractive, but it means working with a bound-state, and therefore departing from ordinary perturbation theory. No one knows how to do this systematically at present, and we do not pursue the idea further here.

6.9   High temperatures and high fermion number densities

In non-relativistic physics, spontaneous symmetry-breaking often disappears as the temperature is raised. This tendency is independent of the presence or absence of a long-range (gauge) force, and is exemplified by ferromagnetism and superconductivity.

The same effect is to be expected in relativistic theories (Kirzhnits

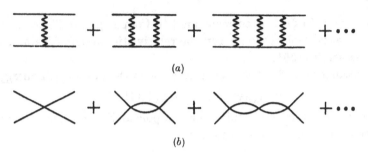

FIGURE 6.1. Graphs summed to generate a Higgs bound-state: (a) with exchange of a vector particle in 4 dimensions, (b) with a contact interaction in 2 dimensions (solid lines are fermions).

and Linde 1972, Weinberg 1974, Dolan and Jackiw 1974, Jacobs 1975). The critical temperature T_c is very roughly given by

$$kT_c \sim fc^2, \tag{6.59}$$

where f is a typical field vacuum-expectation-value. Formally this comes about by a breakdown of perturbation theory (for fixed masses f is of order g^{-1}): some higher-order graphs become large at high temperatures.

In the Weinberg–Salam model of weak interactions (chapter 8), (6.59) gives

$$kT_c \simeq 300\,\mathrm{GeV}. \tag{6.60}$$

A similar effect occurs for high lepton- or baryon-number density (Harrington and Yildiz 1974). The critical chemical potential, μ_c, is

$$\mu_c \sim fc^2. \tag{6.61}$$

Neglecting fermion mass

$$\mu_c \simeq (3\pi^2 n_c)^{\frac{1}{3}}, \tag{6.62}$$

and the critical number density n_c comes out to be

$$n_c \simeq 10^{47}\,\mathrm{cm^{-3}}. \tag{6.63}$$

Temperatures and densities like these might be attained in final stages of gravitational collapse or in the very early universe (see Weinberg 1972: §15.11). Then the spontaneous symmetry-breaking would go away, and (in some sense) an unbroken Yang–Mills theory would be revealed. A number of unanswered questions then come to mind. Is there a long-range force? Classically, in a closed universe, the total charge coupled to a long-range force is zero (there is nowhere for the lines of force to go). Is this relevant? As the universe cools through the critical temperature (or expands through the critical density), can a domain structure of spontaneous symmetry-breaking arise? (See also the end of chapter 17.)

Sufficiently strong magnetic fields may also dispel the symmetry-breaking (Salam and Strathdee 1974).

7

Topology and symmetry-breaking

7.1 Connectedness of group manifolds

There are some interesting ideas – the quantization of charge, the existence of vortex lines and magnetic monopoles – which are related to the topology of a (possibly broken) symmetry group. We begin by recalling one or two mathematical facts (see, for example, Hamermesh 1960: 319; Gilmore 1974: 129).

The group $O(3)$ has a doubly-connected manifold. Its covering group $SU(2)$ is simply-connected. In each case, the manifold is compact. This is typical of many groups used in symmetry-breaking models. (Non-compact groups, like the Lorentz group $O(3, 1)$ cannot be used as internal symmetry groups in any straightforward way because their unitary representations are infinite-dimensional.)

The group $U(1)$, on which electromagnetic gauge-invariance is based, is different. Its manifold, the unit circle, is compact. But it is multiply-connected, because a closed path going a certain number of times around the circle cannot be deformed into one going a different number of times around. The manifold of the covering group is the real line and is not compact. Therefore $U(1)$ behaves qualitatively differently from groups with compact covering groups. So also do product groups like $U(1) \times SU(2)$ (which is the group to be used in chapter 8).

7.2 Charge quantization

It is a fact of observation that all known particles have electric charges which are integral multiples of the electron's charge e. (If quarks were discovered, e might have to be replaced by $\frac{1}{3}e$ in this statement.) There is no fundamental cause in conventional electrodynamics why this should be so. Under a gauge transformation

$$\mathscr{A}_\lambda \to \mathscr{A}_\lambda + \partial_\lambda \omega \tag{7.1}$$

there could be fields ϕ and χ transforming as

$$\phi \to e^{ie\omega}\phi, \quad \chi \to e^{ie'\omega}\chi, \tag{7.2}$$

where the ratio e'/e is not rational. (We made use of this possibility in §6.4.)

In the case of a non-abelian local symmetry group G with a compact covering group, however, all charges are quantized in terms of a finite set of coupling constants g^α (a different one for each factor group of G if G is a direct product). To see this, note that the g^α occur in the transformation law (4.4) of the Yang–Mills fields themselves. Therefore the same g^α must occur in the transformation law of any field that transforms under the group, for any such field couples to the Yang–Mills field. A more rigorous proof has been given by Georgi and Glashow (1974) in footnote 10.

This constitutes a persuasive theoretical argument in favour of the use of groups with compact covering groups. Unfortunately the simplest model with any physical application is based on $SU(2) \times U(1)$ (see chapter 8), and this is not of the desired type. The coupling constant associated with the $U(1)$ subgroup need not be quantized, and the electric charge turns out to be a function of this constant. Perhaps $SU(2) \times U(1)$ is just a subgroup of the true symmetry group.

7.3   Vortex lines

Quantized vortex lines (or thin tubes of magnetic flux) occur in super-fluid helium and in type II superconductors (see, for example, Vinen 1968; Rose-Innes and Rhoderick 1969: §12.2). The analogy between simplified models of superconductors and Higgs models suggests that vortices of some kind might occur in the latter case also. The following argument supports this idea.

Take the abelian Higgs model of §6.2. The quantum theory of spontaneous symmetry-breaking starts from the trivial classical solution

$$\phi(x) = f, \quad \mathscr{A}_\lambda(x) = 0. \tag{7.3}$$

Instead, we seek a less trivial type of classical solution in which, asymptotically,

$$|\phi| \to f, \quad (\partial_\lambda - ie\mathscr{A}_\lambda)\phi \to 0 \quad \text{as} \quad |\mathbf{x}| \to \infty. \tag{7.4}$$

This satisfies asymptotically the classical equations of motion for ϕ derived from (6.8), provided that f is given by (5.6).

More specifically, seek a solution (Nielsen and Olesen 1973) in which, in terms of cylindrical co-ordinates (z, ρ, ψ),

$$\phi \sim f e^{in\psi} \quad \text{as} \quad \rho \to \infty, \tag{7.5}$$

where n is an integer. Then (7.4) is satisfied if

$$\mathscr{A}_\lambda \sim e^{-1} \partial_\lambda \omega, \quad \text{where} \quad \omega = n\psi. \tag{7.6}$$

There is no electromagnetic field in the asymptotic region but the multi-valued nature of the gauge function ω in (7.6) shows that there are flux lines in the neighbourhood of the z-axis.

The state of affairs exemplified here is characterized more generally as follows. From the definition of the current (6.48),

$$j_\lambda = 2\phi^\dagger \phi [e\mathscr{A}_\lambda - \partial_\lambda(\arg \phi)]. \tag{7.7}$$

Therefore the quantity (called the fluxoid in superconductivity)

$$\oint_C [e\mathscr{A}_\lambda - j_\lambda/(2\phi^\dagger\phi)] \, dx^\lambda = 2\pi n, \tag{7.8}$$

where n is an integer (assuming ϕ to be single-valued). In the above example, n is non-zero for any closed curve C encircling the z-axis.

By adjusting the parameters in the potential V in (5.3), the lateral extension of the vortex can be made small. The vortex line may then resemble the 'relativistic string' postulated in dual resonance model theory (Schwarz 1973, Frampton 1974). The excitations of the relativistic string are known to resemble qualitatively the observed pattern of Regge trajectories.

The nature of the vortices (7.6) depends upon the multi-connectedness of $U(1)$. In the case of say $O(3)$, the vortices are not oriented. Any two vortices can annihilate one-another, because the group manifold is only doubly-connected. For possible applications of these ideas to quark models of hadrons, see Nambu (1974), Mandelstam (1975).

7.4 Magnetic monopoles

In conventional electromagnetism, magnetic monopoles can be introduced only with great difficulty (Dirac 1948, Schwinger 1966). The existence of a vector potential $\mathscr{A}_i$ forces the magnetic flux through any closed surface to be zero. The only way out is for $\mathscr{A}_i$ to be many-valued, and to be singular along a curve terminating at the monopole.

Such a curve must join two opposite monopoles, or stretch from one monopole to infinity.

If electromagnetic gauge-invariance is the remnant of the spontaneous breaking of a larger group, other possibilities arise ('t Hooft 1974). As an example, take the group $O(3)$ with a Yang–Mills field W_λ^α and a triplet scalar field ϕ^α. Normally under spontaneous symmetry-breaking one would have $\langle 0|\phi^\alpha|0\rangle = F^\alpha$, and the rotations about F^α would be unbroken and identified with the electromagnetic gauge group.

Analogously to (7.4), we now look for a classical solution in which $(\phi^\alpha)^2$ (but not ϕ^α) is asymptotically constant, and the covariant derivative vanishes:

$$D_\lambda \phi^\alpha \equiv \partial_\lambda \phi^\alpha + g\epsilon^{\alpha\beta\gamma} W_\lambda^\beta \phi^\gamma \sim 0. \tag{7.9}$$

These conditions are satisfied by

$$\left. \begin{aligned} \phi^\alpha &\sim fr^{-1}\delta^{\alpha i}x_i, \\ W_i^\alpha &\sim -\epsilon^{i\alpha j}x_j/(gr^2), \quad W_0^\alpha \sim 0, \end{aligned} \right\} \tag{7.10}$$

where f is a constant. Here $i = 1, 2, 3$ is used to stand for the space values of a 4-vector, and $r^2 = x_i^2$. The Greek and Latin suffices in (7.10) are meant to emphasize the connection between the internal symmetry space and real space which is established. Of course, any *constant* rotation of the internal symmetry group applied to (7.10) would provide an equivalent solution. In this case the field is not zero, but

$$F_{ij}^\alpha \sim -\epsilon_{ijk}x^k x^\alpha/(gr^4), \quad F_{0i}^\alpha = 0. \tag{7.11}$$

The next question is how to define the electromagnetic field. 't Hooft (1974) defines

$$\mathscr{A}_\lambda = (\phi^2)^{-\frac{1}{2}} \phi^\alpha W_\lambda^\alpha, \tag{7.12}$$

$$F_{\lambda\nu} = (\phi^2)^{-\frac{1}{2}} \phi^\alpha F_{\lambda\nu}^\alpha - g^{-1}(\phi^2)^{-\frac{3}{2}} \epsilon^{\alpha\beta\gamma}\phi^\alpha (D_\lambda \phi^\beta)(D_\nu \phi^\gamma) \tag{7.13}$$

(the covariant derivation $D_\lambda \phi^\beta$ is defined in (7.9)). If ϕ were a con-constant, say $(0, 0, f)$, these equations would reduce to

$$\mathscr{A}_\lambda = W_\lambda^3, \quad F_{\lambda\nu} = \partial_\lambda W_\nu^3 - \partial_\nu W_\lambda^3 \tag{7.14}$$

(note the cancellation between pieces quadratic in W from the two terms on the right of (7.13)). In general, however, the relation between (7.12) and (7.13) is more complicated:

$$F_{\lambda\nu} = \partial_\lambda \mathscr{A}_\nu - \partial_\nu \mathscr{A}_\lambda - g^{-1}(\phi^2)^{-\frac{3}{2}} \epsilon^{\alpha\beta\gamma}\phi^\alpha (\partial_\lambda \phi^\beta)(\partial_\nu \phi^\gamma). \tag{7.15}$$

In the present example, insertion of (7.10) into (7.12) and (7.13) yields

$$\mathscr{A}_\lambda \sim 0, \quad F_{ij} \sim -g^{-1}\epsilon_{ijk}x_k/r^3, \quad \text{as} \quad r \to \infty. \tag{7.16}$$

The latter expression is recognized as the magnetic field of a monopole of strength g^{-1}. It is not produced by an ordinary vector potential, but by the second term in (7.15). 't Hooft (1974) has estimated the mass of such a monopole and finds it to be of the order

$$4\pi g^{-2}M, \tag{7.17}$$

where M is a typical vector-meson mass in the model. If $g \simeq e$, (7.17) gives an enormous mass of order $137M$.

In this theory, a magnetic monopole is a point about which the direction of the local value of $\phi^\alpha(x)$ swings. If two opposite monopoles are placed near together, the value of $\phi^\alpha(x)$ is asymptotically constant in all directions. Only near the monopoles does it vary; and directly in between it has the opposite direction to the asymptotic one.

It should be emphasized that in §7.3 and in this section we have described solutions of the *classical* equations. The problem of incorporating these ideas into quantum theory is a very difficult one.

8

A theory of leptons

8.1 The model

In this chapter we describe the theory of the interactions (weak and electromagnetic) of leptons that is due to the (independent) work of Weinberg (1967) and Salam (1968). There have been other attempts to unify weak and electromagnetic interactions (Schwinger 1957, Glashow 1961, Salam and Ward 1964, T. D. Lee 1971). The new feature of Salam's and Weinberg's model was that it was based on Higgs' mechanism. They each conjectured that the resulting theory might be renormalizable. We have already, §6.3, given arguments to suggest that they were correct, and we will return to this in chapter 14. In this chapter and the next, we will concentrate on the immediate experimental implications.

In this original Salam–Weinberg theory, the muon and its neutrino are treated in exactly the same way as the electron and its neutrino.

Let the field of the negative electron (muon) be $e(\mu)$ and of its neutrino be $\nu_e(\nu_\mu)$. Define the left- and right-handed components

$$e_{\mathrm{L,R}} = \tfrac{1}{2}(1 \pm \gamma_5)\, e, \quad \mu_{\mathrm{L,R}} = \tfrac{1}{2}(1 \pm \gamma_5)\, \mu.$$

In order to induce the correct weak coupling to the charged W-mesons (considered as Yang–Mills fields), it is natural to put the left-handed fields into doublets

$$l_e = \begin{pmatrix} \nu_e \\ e_{\mathrm{L}} \end{pmatrix}, \quad l_\mu = \begin{pmatrix} \nu_\mu \\ \mu_{\mathrm{L}} \end{pmatrix}, \tag{8.1}$$

which transform as spinor representations of the weak group $SU(2)_{\mathrm{L}}$ whose hadronic part was defined in §2.2. But the neutral member of the triplet of Yang–Mills fields certainly cannot be identified with the photon, since it couples to the neutrinos in (8.1) and it does not couple to e_{R} and μ_{R}. To introduce the photon one may proceed as follows.

Let us term $SU(2)_{\mathrm{L}}$ the 'weak iso-spin group', and call the corresponding quantum number **t**. Define a 'weak hypercharge' y by writing the electric charge operator as

$$Q = t_3 + \tfrac{1}{2}y. \tag{8.2}$$

Then the doublets l_e and l_μ must have $y = -1$. Since e_R and μ_R each apparently have no leptonic partner, we assign each of them to a singlet and they must have $y = -2$. These quantum numbers are summarized in table 8.1. The bottom row is for the Higgs field, to be introduced shortly.

Before spontaneous symmetry-breaking, the leptons are massless, since mass terms would not commute with the $(1 + \gamma_5)$ in $SU(2)_L$.

TABLE 8.1   *Quantum numbers in the Salam–Weinberg model*

Particle	t	t_3	y	Q
ν_e, ν_μ	$\tfrac{1}{2}$	$\tfrac{1}{2}$	-1	0
e_L, μ_L	$\tfrac{1}{2}$	$-\tfrac{1}{2}$	-1	-1
e_R, μ_R	0	0	-2	-1
ϕ	$\tfrac{1}{2}$	$\pm\tfrac{1}{2}$	1	$1, 0$

8.2   The vector fields

It is natural to connect y with a local gauge group $U(1)$, associated with a vector field B_λ. The complete local gauge group G of the weak interactions is now $SU(2)_L \times U(1)_y$. Let the coupling constants of $\mathbf{W}_\lambda$ and B_λ be g and g' respectively. (Strictly, for $U(1)$, there is no *a priori* reason why a different coupling should not be associated with each representation. This was explained in §7.1.) Suppose now that the symmetry is spontaneously broken in such a way that the subgroup $U(1)^{em}$, generated by the operator Q in (8.2), remains an unbroken local gauge symmetry. The massless field associated with $U(1)^{em}$ is interpreted as the photon field $\mathscr{A}_\lambda$, which is some linear combination of W_λ^3 and B_λ. The orthogonal combination, which we call Z_λ, represents a neutral, massive spin-1 particle; and W_λ^1, W_λ^2 represent massive charged spin-1 particles.

We describe this mixing of the neutral fields by a mixing-angle θ_W (whose value will be determined shortly):

$$\left. \begin{aligned} B_\lambda &= \cos\theta_W \, \mathscr{A}_\lambda - \sin\theta_W Z_\lambda, \\ W_\lambda^3 &= \sin\theta_W \, \mathscr{A}_\lambda + \cos\theta_W Z_\lambda. \end{aligned} \right\} \tag{8.3}$$

The coupling of the vector mesons to leptons has the form

$$g\mathbf{j}_\lambda \cdot \mathbf{W}^\lambda + \tfrac{1}{2}g'j_\lambda^y B^\lambda, \tag{8.4}$$

where $\qquad \int \mathrm{d}^3 x \mathbf{j}_0 = \mathbf{t}, \quad \int \mathrm{d}^3 \mathbf{x} j_0^y = y.$ (8.5)

Using (8.3), (8.4) becomes

$$g(j_\lambda^1 W^{1\lambda} + j_\lambda^2 W^{2\lambda}) + (g \sin \theta_W j_\lambda^3 + \tfrac{1}{2} g' \cos \theta_W j_\lambda^y) \mathscr{A}^\lambda$$
$$+ (g \cos \theta_W j_\lambda^3 - \tfrac{1}{2} g' \sin \theta_W j_\lambda^y) Z^\lambda. \quad (8.6)$$

By (8.2) and (8.5), the electromagnetic interaction is

$$e j_\lambda^{\mathrm{em}} \mathscr{A}^\lambda = e(j_\lambda^3 + \tfrac{1}{2} j_\lambda^y) \mathscr{A}^\lambda. \quad (8.7)$$

Comparison of (8.6) with (8.7) gives

$$e = g \sin \theta_W = g' \cos \theta_W \quad (8.8)$$

or $\qquad \tan \theta_W = g'/g, \quad e = g g' (g^2 + g'^2)^{-\frac{1}{2}}.$ (8.9)

Using these equations, (8.6) may be written

$$e j_\lambda^{\mathrm{em}} \mathscr{A}^\lambda + 2^{-\frac{1}{2}} g (j^\lambda W_\lambda^\dagger + j_\lambda^\dagger W^\lambda) + g (\cos \theta_W)^{-1} j_\lambda^Z Z^\lambda, \quad (8.10)$$

where $\qquad j_\lambda = j_\lambda^1 + i j_\lambda^2, \quad W_\lambda = 2^{-\frac{1}{2}} (W_\lambda^1 + i W_\lambda^2) \Big\}$
and $\qquad j_\lambda^Z = \cos^2 \theta_W j_\lambda^3 - \tfrac{1}{2} \sin^2 \theta_W j_\lambda^y. \Big\}$ (8.11)

The charged current j_λ is given by (2.9). Of course, (8.10) contains the conventional lepton interaction in (2.7).

From (8.8) follows the inequality

$$g > e. \quad (8.12)$$

Also, (2.1), (2.25) and (8.8) give

$$M_W = 2^{-\frac{1}{4}} G_W^{-\frac{1}{2}} e / \sin \theta_W \simeq (37 / \sin \theta_W) \, \mathrm{GeV}. \quad (8.13)$$

The apparent weakness of weak interactions is mainly due to the large mass of the W-meson. A particle with mass greater than 37 GeV, as required by (8.13), would not have been detected yet.

Finally, with the aid of table 8.1, we write out the Z current (8.11) in terms of the lepton fields:

$$j_\lambda^Z = \tfrac{1}{2} \bar{\nu}_e \gamma_\lambda \nu_e - \tfrac{1}{2} \cos 2\theta_W \, \bar{e}_L \gamma_\lambda e_L + \sin^2 \theta_W \bar{e}_R \gamma_\lambda e_R$$
$$+ (\text{terms with } e \to \mu). \quad (8.14)$$

8.3   The Higgs field

So far, the model has yielded the lower limit on M_W and predicted the existence of the neutral current j_λ^Z (together with its general form). But the parameters M_Z and θ_W remain undetermined and therefore also the effective strength of the neutral current interaction (proportional to M_Z^{-2} at not too high energies).

Consider the Higgs field ϕ whose vacuum-expectation-value produces the symmetry-breaking. Suppose ϕ is a multiplet with weak iso-spin t', and weak hypercharge y'.

If $\langle 0|\phi|0\rangle = F$, then F must be electrically neutral; and so, by (8.2), have

$$t_3' = -\tfrac{1}{2}y'. \tag{8.15}$$

The W-mass is given by (6.30) (generalized to complex representations) to be

$$\tfrac{1}{2}M_W^2 = g^2 F^\dagger t_1^2 F = \tfrac{1}{2}g^2[t'(t'+1)-t_3'^2]\,|F|^2. \tag{8.16}$$

Similarly, using (8.3), (8.8) and (8.15),

$$\tfrac{1}{2}M_Z^2 = F^\dagger(\cos\theta_W\, g t_3' - \tfrac{1}{2}\sin\theta_W\, g'\, y')^2\, F$$

$$= (g^2+g'^2)\, t_3'^2|F|^2. \tag{8.17}$$

These equations place no restriction on the ratio M_Z/M_W; so to make further progress we must make some assumption about t' or y'.

As we shall see in §8.4, ϕ has to couple to $\bar{e}_L e_R$ and $\bar{\mu}_L \mu_R$ in order to provide lepton masses. This is possible if $t' = \tfrac{1}{2}$, $y' = 1$, and we now *assume* ϕ to be irreducible with these quantum numbers. Equations (8.16) and (8.17) then give

$$M_W^2 = \tfrac{1}{2}g^2|F|^2, \quad M_Z^2 = \tfrac{1}{2}(g^2+g'^2)|F|^2 \tag{8.18}$$

or $\qquad\qquad M_Z = M_W(\cos\theta_W)^{-1} \simeq (74/\sin 2\theta_W)\,\mathrm{GeV}. \tag{8.19}$

Thus the two unknown parameters are reduced to one. It should however be stressed that the assumption that ϕ transforms as a $t = \tfrac{1}{2}$ irreducible representation could well be wrong.

As in (6.29), we write

$$\phi = F + \phi' \equiv 2^{-\frac{1}{2}}\binom{0}{f} + \phi', \tag{8.20}$$

where F has $t_3' = -\tfrac{1}{2}$ as required by (8.15). There is no loss of generality in this choice of F. Any constant spinor can be brought into this form, and a different choice would just mean a re-labelling of the lepton and vector-meson field components.

It is convenient to write out ϕ' in the form

$$\phi' = 2^{-\frac{1}{2}}(\chi' + i\boldsymbol{\tau}\cdot\boldsymbol{\varphi}')\binom{0}{1} = 2^{-\frac{1}{2}}\binom{\phi_2'+i\phi_1'}{\chi'-i\phi_3'}, \tag{8.21}$$

where, in spite of the notation, $\boldsymbol{\varphi}'$ does not transform as a vector under $SU(2)_L$. Let $\boldsymbol{\omega}$ and ω be infinitesimal parameters of the $SU(2)_L$ and $U(1)_y$ groups. Under such a transformation,

$$\phi' \to \phi' + \tfrac{1}{2}i(g\boldsymbol{\tau}\cdot\boldsymbol{\omega}+g'\omega)(F+\phi'). \qquad (8.22)$$

Therefore, using (8.21), χ transforms homogeneously but $\boldsymbol{\varphi}'$ acquires an inhomogeneous term

$$\tfrac{1}{2}f(g\omega_1,g\omega_2,g\omega_3-g'\omega). \qquad (8.23)$$

This is why (8.21) is a convenient notation. χ' represents a genuine, neutral spinless Higgs particle, but ϕ_1', ϕ_2', ϕ_3' have no physical significance independently of the vector fields W_λ^1, W_λ^2, Z_λ.

8.4 The complete Lagrangian

Knowing the structure of the theory, we can now write out the complete Lagrangian density in the form

$$\mathscr{L} = \mathscr{L}_W + \mathscr{L}_B + \mathscr{L}_l + \mathscr{L}_\phi + \mathscr{L}_{\phi l} - V + \mathscr{L}_G. \qquad (8.24)$$

$\mathscr{L}_W$ is the Yang–Mills Lagrangian (4.9), and $\mathscr{L}_B$ is

$$\mathscr{L}_B = -\tfrac{1}{2}(\partial_\lambda B_\nu - \partial_\nu B_\lambda)\partial^\lambda B^\nu \qquad (8.25)$$

like the electromagnetic Lagrangian (3.20). The invariant lepton Lagrangian is (reading off the couplings from table 8.1)

$$\mathscr{L}_l = i\bar{l}_e\gamma^\lambda(\partial_\lambda - \tfrac{1}{2}ig\boldsymbol{\tau}\cdot\mathbf{W}_\lambda + \tfrac{1}{2}ig'B_\lambda)l_e$$
$$+ i\bar{e}_R\gamma^\lambda(\partial_\lambda + ig'B_\lambda)e_R + (e\leftrightarrow\mu). \qquad (8.26)$$

Similarly, $\qquad \mathscr{L}_\phi = |(\partial_\lambda - \tfrac{1}{2}ig\boldsymbol{\tau}\cdot\mathbf{W}_\lambda - \tfrac{1}{2}ig'B_\lambda)\phi|^2. \qquad (8.27)$

The Higgs fields can be coupled invariantly to the leptons:

$$\mathscr{L}_{\phi l} = -2^{\frac{1}{2}}m_e f^{-1}[\bar{e}_R(\phi^\dagger l_e) + (\bar{l}_e\phi)e_R] + (e\leftrightarrow\mu). \qquad (8.28)$$

The coupling strength has been chosen so that the F term in (8.20) contributes the correct mass terms in (8.28) (remember that

$$\bar{e}_R e_L + \bar{e}_L e_R = \bar{e}e).$$

The $t = \tfrac{1}{2}$, $y' = 1$ representation for ϕ was chosen with (8.28) in view. The self-interaction can be written, like (5.3),

$$V = \lambda(\phi^\dagger\phi - \tfrac{1}{2}f^2)^2, \qquad (8.29)$$

where λ is an arbitrary positive constant. ϕ' is defined in (8.20) so that $\phi' = 0$ is a minimum of (8.29).

If (8.20) is put into (8.27), the masses (8.18) are of course obtained,

$$M_W = \tfrac{1}{2}fg, \quad M_Z = \tfrac{1}{2}f(g^2 + g'^2)^{\frac{1}{2}}. \tag{8.30}$$

The magnitude of f is fixed by (2.1) and (2.25)

$$f = 2^{-\frac{1}{4}}G_W^{-\frac{1}{2}} \simeq 246 \, \text{GeV}. \tag{8.31}$$

The coupling constants in (8.28) are very small:

$$\left.\begin{aligned}
2^{\frac{1}{4}}m_e f^{-1} &= 2^{\frac{3}{4}}m_e G_W^{\frac{1}{2}} \simeq 3 \times 10^{-6}, \\
2^{\frac{1}{4}}m_\mu f^{-1} &= 2^{\frac{3}{4}}m_\mu G_W^{\frac{1}{2}} \simeq 6 \times 10^{-4}.
\end{aligned}\right\} \tag{8.32}$$

It is an unsatisfactory feature of the model, that these numbers have to be arbitrarily inserted into $\mathscr{L}$. V in (8.29) gives the χ-field a mass

$$M_\chi^2 = 2\lambda f^2. \tag{8.33}$$

Neither M_χ nor λ are known at present. Limits on M_χ will be mentioned in §9.5.

Finally, before one can calculate with the Lagrangian, a gauge-fixing term is needed. In conformity with (6.32) (and generalizing (6.17)) we take for $\mathscr{L}_G$ in (8.24) the 't Hooft term

$$\begin{aligned}
\mathscr{L}_G &= -\tfrac{1}{2}\xi^{-1}\left[\sum_{\alpha=1}^{3}(\partial\cdot W^\alpha + \xi M_W \phi'_\alpha)^2 + (\partial\cdot B - \tfrac{1}{2}\xi fg'\phi'_3)^2\right] \\
&= -\tfrac{1}{2}\xi^{-1}\left[\sum_{\alpha=1}^{2}(\partial\cdot W^\alpha + \xi M_W \phi'_\alpha)^2 + (\partial\cdot Z + \xi M_Z \phi'_3)^2 + (\partial\cdot\mathscr{A})^2\right].
\end{aligned} \tag{8.34}$$

Here ϕ'_i ($i = 1, 2, 3$) is defined in (8.21). The W-ϕ and B-ϕ mixing terms in (8.27) should be cancelled by those in (8.34). To verify this, write the mixing terms in (8.27) as

$$\tfrac{1}{2}\mathrm{i}F^+(g\boldsymbol{\tau}\cdot\mathbf{W}_\lambda + g'B_\lambda)\,\partial^\lambda\phi' + (\text{Herm. conj.}),$$

and use the first form for ϕ' in (8.21). To check that the two forms of (8.34) are equal use (8.3), (8.8) and (8.18). Note that the second form gives the 'masses' of ϕ'_1, ϕ'_2, ϕ'_3 as $\xi^{\frac{1}{2}}M_W$, $\xi^{\frac{1}{2}}M_W$, $\xi^{\frac{1}{2}}M_Z$.

8.5 Neutrino scattering on electrons

The most easily tested prediction of the Salam–Weinberg model is the existence of the neutral vector meson Z, with lepton current (8.11) and mass limited by (8.19) (the latter assuming the simplest choice of

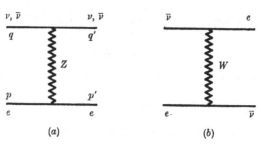

FIGURE 8.1. Neutrino-electron scattering graphs.

Higgs field). We discuss the effect of Z-exchange on three neutrino reactions:

$$\nu_\mu e^- \to \nu_\mu e^-, \tag{8.35}$$

$$\bar{\nu}_\mu e^- \to \bar{\nu}_\mu e^-, \tag{8.36}$$

$$\bar{\nu}_e e^- \to \bar{\nu}_e e^-. \tag{8.37}$$

The first two proceed solely by Z-exchange. The third has contributions from a W-pole also, as shown in fig. 8.1. The exchange of the ϕ_i' fields in Born approximation can be neglected, because the coupling to leptons is so small in (8.28) and (8.32). To the same approximation, the $k_\lambda k_\nu$ terms in the propagator (6.19) can be neglected (they give lepton masses because of the Dirac equations satisfied by the initial and final lepton spinors); and so the choice of ξ is immaterial.

Neglecting the momentum transfer compared with M_W or M_Z in the propagator, the effective interaction is, by (8.10)

$$-\tfrac{1}{2}g^2(M_Z\cos\theta_W)^{-2}j_\lambda^Z j^{Z\lambda} - \tfrac{1}{2}g^2 M_W^{-2} j_\lambda^\dagger j^\lambda \tag{8.38}$$

or, by (8.19), (8.30) and (8.31),

$$-2^{\frac{1}{2}}G_W(j_\lambda^Z j^{Z\lambda} + j_\lambda^\dagger j^\lambda). \tag{8.39}$$

(The minus sign comes from factors of i in vertex and propagator.) Substituting from (8.14), the relevant part of the interaction (8.39) is

$$2^{\frac{1}{2}}G_W\{\bar{\nu}_\mu\gamma^\lambda\tfrac{1}{2}(1+\gamma_5)\nu_\mu + \bar{\nu}_e\gamma^\lambda\tfrac{1}{2}(1+\gamma_5)\nu_e\}$$
$$\times\{\bar{e}\gamma_\lambda[\tfrac{1}{2}(1+\gamma_5)\cos2\theta_W - (1-\gamma_5)\sin^2\theta_W]e\}$$
$$-2^{\frac{1}{2}}G_W\{\bar{e}\gamma^\lambda\tfrac{1}{2}(1+\gamma_5)\nu_e\}\{\bar{\nu}_e\gamma_\lambda\tfrac{1}{2}(1+\gamma_5)e\}, \tag{8.40}$$

where the chiral projection operators $\tfrac{1}{2}(1\pm\gamma_5)$ have been made explicit.

It is convenient to discuss the first term in (8.40) as a special case of the general interaction

$$2^{\frac{1}{2}}G_W\{\bar{\nu}_\mu\gamma^\lambda\tfrac{1}{2}(1+\gamma_5)\nu_\mu\}\{\bar{e}\gamma_\lambda[\tfrac{1}{2}(1+\gamma_5)c_L + \tfrac{1}{2}(1-\gamma_5)c_R]e\}, \tag{8.41}$$

where c_R and c_L are constants. For the process (8.35) the differential cross-section, with respect to the electron recoil energy E'_e in the lab. frame, is found to be ('t Hooft 1971b)

$$\frac{d\sigma}{dE'_e} = [G_W^2/(2\pi m_e E_\nu^2)]\,[\,|c_L|^2\,(p\cdot q)^2 + |c_R|^2\,(p'\cdot q)^2$$
$$+ \tfrac{1}{2}(c_R^* c_L + c_L^* c_R)\,m_e^2 q\cdot q'], \quad (8.42)$$

where p, p' are the initial and final 4-momenta for the electron, q, q' are those for the neutrino, and $q_0 = E_\nu$, $p'_0 = E'_e$. In terms of the ratio $y = E'_e/E_\nu$, (8.42) becomes

$$\frac{d\sigma}{dy} = (G_W^2 m_e E_\nu/2\pi)\,[\,|c_L|^2 + |c_R|^2\,(1-y)^2$$
$$+ \tfrac{1}{2}(c_R^* c_L + c_L^* c_R)\,y m_e/E_\nu]. \quad (8.43)$$

The left-right interference term here is negligible if $E_\nu \gg m_e$. The derivation of (8.42) is standard, using the identity

$$\tfrac{1}{2}\mathrm{Tr}\,[q\cdot\gamma\gamma^\lambda p\cdot\gamma\gamma^\nu \tfrac{1}{2}(1\pm\gamma_5)] = p^\lambda q^\nu + p^\nu q^\lambda - g^{\lambda\nu}p\cdot q \pm i\epsilon^{\lambda\nu\sigma\rho}q_\sigma p_\rho. \quad (8.44)$$

To get the cross-section for the anti-neutrino process (8.36), one interchanges c_L with c_R in (8.43) (because anti-neutrinos have opposite helicity to neutrinos).

The total cross-sections for (8.35) and (8.36) are then found from (8.40), (8.41) and (8.43) to be (neglecting the interference terms)

$$(G_W^2 m_e E_\nu/2\pi)\,[(1 - 2\sin^2\theta_W)^2 + \tfrac{4}{3}\sin^4\theta_W] \quad (8.45)$$

and $\quad (G_W^2 m_e E_\nu/2\pi)\,[\tfrac{1}{3}(1 - 2\sin^2\theta_W)^2 + 4\sin^4\theta_W]. \quad (8.46)$

These functions are shown in fig. 8.2(a) and (b) where the ratio (σ/E_ν) is plotted, in units of $(G_W^2 m_e/2\pi)$, against $\sin^2\theta_W$. Note that

$$G_W^2 m_e/2\pi \simeq 0.4 \times 10^{-41}\,\mathrm{cm^2/GeV}. \quad (8.47)$$

The processes (8.35) and (8.36) have been sought in the Gargamelle heavy-liquid bubble-chamber at CERN (Hasert $et\ al.$ 1973a). The event rate is expected to be roughly 1 per million exposures. In the $\bar\nu$ beam, by March 1975, three events have been reported with electron tracks that are candidates for (8.36). From (8.46) (corrected for the acceptance of E'_e) a limit is deduced

$$\sin^2\theta_W < 0.45. \quad (8.48)$$

Background in this experiment is considered to be low and, if more events are found, it will provide decisive evidence for neutral currents.

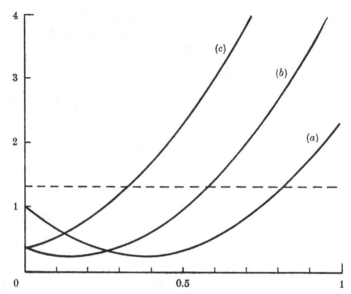

FIGURE 8.2. Neutrino-electron scattering cross-sections in units of $G_W^2 m_e E_\nu / 2\pi$, against $\sin^2 \theta_W$: (a) $\nu_\mu e^-$, (b) $\bar{\nu}_\mu e^-$, (c) $\bar{\nu}_e e^-$ (broken line is W-exchange contribution).

In the scattering of electron-type anti-neutrinos, (8.37), the W-pole graph fig. 8.1 (b) has to be included, and gives the last term of (8.40):

$$-2^{\frac{3}{2}} G_W [\bar{e}\gamma^\lambda \tfrac{1}{2}(1+\gamma_5)\nu_e][\bar{\nu}_e \gamma_\lambda \tfrac{1}{2}(1+\gamma_5)e]. \qquad (8.49)$$

The contribution from fig. 8.1 (a) is given in the second term of (8.40). To combine this with (8.49), it is necessary to do a Fierz transformation (Good 1955) on (8.49), that is to write it with the spinors paired off differently. This is an easy task in the present case. The spinors are all effectively two-component ones (in virtue of the $\tfrac{1}{2}(1+\gamma_5)$ projection operators), and 4-vectors are the only covariants that can be formed from a two-component spinor and the Hermitian conjugate of a two-component spinor. Thus (8.49) must be equal to

$$-2^{\frac{3}{2}} \eta G_W [\bar{\nu}_e \gamma^\lambda \tfrac{1}{2}(1+\gamma_5)\nu_e][\bar{e}\gamma_\lambda \tfrac{1}{2}(1+\gamma_5)e], \qquad (8.50)$$

where only the factor η remains to be determined. Since (8.49) and (8.50) are identically equal, a special case is sufficient to fix η, for example

$$\nu = \begin{pmatrix} a \\ 0 \\ 0 \\ 0 \end{pmatrix}, \quad e = \begin{pmatrix} 0 \\ b \\ 0 \\ 0 \end{pmatrix}.$$

Remembering that $aa^\dagger = -a^\dagger a$, $bb^\dagger = -b^\dagger b$, one finds that

$$\eta = 1. \tag{8.51}$$

Combining (8.50) and (8.40), one obtains (8.41) with

$$c_L = \cos 2\theta_W - 2, \quad c_R = -2\sin^2\theta_W. \tag{8.52}$$

The total cross-section for process (8.37) is then obtained from (8.43) and (8.52), but interchanging c_R with c_L (for anti-neutrino scattering):

$$(G_W^2 m_e E_\nu/2\pi)\,[\tfrac{1}{3}(1 + 2\sin^2\theta_W)^2 + 4\sin^4\theta_W]. \tag{8.53}$$

This function is plotted in fig. 8.2(c).

An upper limit has been placed on this cross-section in an experiment using anti-neutrinos from a fission reactor (Gurr, Reines and Sobel 1974). After suitable weighting with lepton energy distributions appropriate to the experiment, the expected rate is still an increasing function of $\sin^2\theta_W$. The upper limit found for the averaged cross-section was 6×10^{-47} cm^2. This is irrelevant to the existence of neutral currents, since it is consistent with the contribution from W-meson exchange on its own (represented by the horizontal broken line on fig. 8.2). Assuming the Salam–Weinberg theory, the experiment implies

$$\sin^2\theta_W < 0.35. \tag{8.54}$$

On the face of it, this is the best available limit on θ_W, but the experiment is a difficult one, relying on differences in counting rate with the reactor on or off. (See Perkins (1974) for a discussion of the technical problems in this and other neutrino experiments.) Accepting (8.54), (8.13) gives

$$M_W \geqslant 64\,\text{GeV}. \tag{8.55}$$

8.6   Electron–positron annihilation

Another way in which the influence of the Z-meson might show up in future experiments is in its interference with the photon in

$$e^+e^- \to \mu^+\mu^-. \tag{8.56}$$

As the total centre-of-mass energy E rises the contribution from the Z-meson increases because of the propagator $(M_Z^2 - E^2)^{-1}$. Budny (1973) calculated that, for $E = 20\,\text{GeV}$, parity violation effects, including muon longitudinal polarization and muon asymmetry from polarized electrons, can approach 10 %. Multi-photon processes may make the analysis more difficult.

8.7   Special values of θ_W

In the basic Salam–Weinberg model, as described in this chapter, θ_W is a free parameter. Are there any models capable of predicting its value?

A simple remark is that, for $\theta_W = 30°$, the Z-current (8.14) of electrons and muons become pure axial vector. It is conceivable that, for some reason at present unknown, nature requires this to happen. The angle 30° is indeed predicted by a more sophisticated model of Weinberg's (§18.7(i)). It is consistent with present experimental information ((8.48), (8.54), (9.33)). Another speculative model (§18.7(ii)) predicts a numerically similar angle.

8.8   Heavy leptons

In order to fill representations of larger groups than $SU(2) \times U(1)$, some models have postulated the existence of so far undiscovered heavy leptons. One striking possibility is a particle with lepton number and charge of the same sign. A number of such models are discussed by Bjorken and Llewellyn Smith (1973). For experimental lower limits on masses of heavy leptons, see Eichten *et al.* (1973), B. C. Barish *et al.* (1974), Asratyan *et al.* (1974).

9

A provisional model of hadron weak interactions

9.1 Difficulty with strangeness

The strangeness quantum number makes it difficult to extend the theory of chapter 8 to hadrons. Unlike non-strange neutral weak currents, strange neutral currents have stringent limits from decay branching ratios, as illustrated in table 9.1 (see also the particle data tables, Chaloupka *et al.* 1974). (These decays are expected to occur by higher-order processes. For example, the chain

$$K_L^0 \to \pi^+\pi^- \to 2\gamma \to \mu^+\mu^-$$

leads to a predicted lower limit of 0.6×10^{-9} for the branching ratio, see Sehgal 1969.)

TABLE 9.1   *Rare decay modes involving neutral currents and change of strangeness (from Gaillard and Lee 1974b)*

Decay mode	Branching ratio
$K^+ \to \pi^+\mu^+\mu^-$	$< 5.6 \times 10^{-7}$
$K^+ \to \pi^+e^+e^-$	$< 4 \times 10^{-7}$
$K_L^0 \to \mu^+\mu^-$	$\simeq 10^{-8}$

On the other hand, there is an argument to show that the most obvious gauge models for hadrons require strange neutral currents. The argument is simplest expressed in terms of quarks (see table 2.1). The process shown in fig. 9.1 (*a*) occurs from the ordinary weak current (2.20). Like fig. 2.1, it leads to an unacceptable high-energy behaviour which, in analogy with the leptons, one would expect to be cancelled (mainly) by fig. 9.1(*b*). But this contains a neutral vector meson (not the photon, because of parity violation) coupled to the strange neutral current $(\bar{\lambda}n)$. There is no way out of this difficulty without complicating the model.

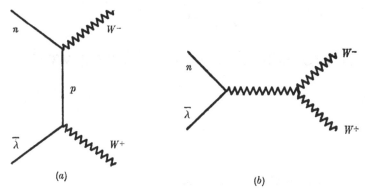

FIGURE 9.1. Strangeness-changing quark diagrams.

Put more formally, the argument runs as follows. To accommodate hadrons in the Salam–Weinberg group $SU(2)_\mathrm{L} \times U(1)_y$, the leptonic currents j^a_λ, j^y_λ defined in § 8.2 have to have hadronic parts added, which we call J^a_λ, J^y_λ. In order to get the ordinary charged weak current (2.21) right, it is natural to try the identification.

$$\left.\begin{aligned} J^a_\lambda &= L'^a_\lambda \quad (a = 1, 2, 3), \\ \tfrac{1}{2} J^y_\lambda &= J^\mathrm{em}_\lambda - L'^3_\lambda, \end{aligned}\right\} \tag{9.1}$$

where L'^a_λ was defined in (2.17). But

$$L'^3_\lambda = (1 - \tfrac{1}{2}\sin^2\theta_\mathrm{C}) L^3_\lambda + \frac{\sqrt{3}}{2}\sin^2\theta_\mathrm{C} L^8_\lambda - \cos\theta_\mathrm{C}\sin\theta_\mathrm{C} L^6_\lambda, \tag{9.2}$$

and L^6_λ has non-zero strangeness.

9.2 A provisional model

Probably the simplest scheme (Glashow, Iliopoulos, Maiani 1970) for circumventing this difficulty postulates a new quantum number C (for 'charm'), with all presently known particles having $C = 0$ (just as 30 years ago all known particles had zero strangeness). The scheme is most simply expressed in terms of quarks. In addition to p, n, λ, which have $C = 0$, imagine a fourth quark c with $C = 1$, $Y = 0$, $Q = \tfrac{2}{3}$, $T = 0$. Let the left-handed doublets

$$\begin{pmatrix} p_\mathrm{L} \\ n'_\mathrm{L} \end{pmatrix}, \quad \begin{pmatrix} c_\mathrm{L} \\ \lambda'_\mathrm{L} \end{pmatrix}. \tag{9.3}$$

be spinor representations of the weak group $SU(2)_\mathrm{L}$, and the right-handed components $p_\mathrm{R}, n'_\mathrm{R}, c_\mathrm{R}, \lambda'_\mathrm{R}$ each be singlets. Assign values of the weak hypercharge y to give the electric charges correctly according to (8.2). Then the neutral current J_λ^3 is

$$J_\lambda^3 = \tfrac{1}{2}(\bar{p}_L \gamma_\lambda p_L - \bar{n}'_L \gamma_\lambda n'_L + \bar{c}_L \gamma_\lambda c_L - \bar{\lambda}'_L \gamma_\lambda \lambda'_L)$$

$$= \tfrac{1}{2}(\bar{p}_L \gamma_\lambda p_L - \bar{n}_L \gamma_\lambda n_L + \bar{c}_L \gamma_\lambda c_L - \bar{\lambda}_L \gamma_\lambda \lambda_L), \tag{9.4}$$

which has zero strangeness (and is actually independent of the Cabibbo angle). The charged currents $J_\lambda^{1,2}$ have charm-changing parts. The bad high-energy behaviour of fig. 9.1 (a) is cancelled, not by (b), but by a graph like (a) in which a c-quark is exchanged.

We may now extract the essential properties of the weak currents in this model. First write

$$\left. \begin{aligned} J_\lambda^{1,2} &= L_\lambda'^{1,2} + \text{(charmed part)}, \\ J_\lambda^3 &= L_\lambda^3 + \tfrac{1}{2}L_\lambda^C - \tfrac{1}{2}L_\lambda^S, \end{aligned} \right\} \tag{9.5}$$

where L_λ^S and L_λ^C are the left-handed strangeness and charm currents. Note that the Cabibbo angle occurs in $J_\lambda^{1,2}$ but not in J_λ^3. We can then obtain the hadronic neutral current J_λ^Z by analogy with (8.11), and write it in a number of alternative ways:

$$J_\lambda^Z = \cos^2\theta_W J_\lambda^3 - \tfrac{1}{2}\sin^2\theta_W J_\lambda^Y$$

$$= \cos^2\theta_W J_\lambda^3 - \sin^2\theta_W (J_\lambda^{em} - J_\lambda^3)$$

$$= J_\lambda^3 - \sin^2\theta_W J_\lambda^{em}, \tag{9.6}$$

$$= L_\lambda^3 - \sin^2\theta_W J_\lambda^{em} + \tfrac{1}{2}L_\lambda^C - \tfrac{1}{2}L_\lambda^S, \tag{9.7}$$

$$= \tfrac{1}{2}(1 - 2\sin^2\theta_W) V_\lambda^3 + \tfrac{1}{2}A_\lambda^3 - \tfrac{1}{2}\sin^2\theta_W V_\lambda^Y$$

$$\quad + \text{(iso-singlets with } S \neq 0 \text{ or } C \neq 0), \tag{9.8}$$

$$= \tfrac{1}{2}(1 - 2\sin^2\theta_W) V_\lambda^3 + \tfrac{1}{2}A_\lambda^3 + \text{(iso-singlets)}. \tag{9.9}$$

Similarly, the charged currents are

$$J_\lambda^{1,2} = L_\lambda'^{1,2} + \text{(charmed pieces)}$$

$$\simeq \tfrac{1}{2}(V_\lambda^{1,2} + A_\lambda^{1,2}) + \text{(charmed pieces)}, \tag{9.10}$$

with neglect of the Cabibbo angle θ_C in the latter form. Both (9.7) and (9.10) contain contributions about which little is known except that they are iso-singlets, and these contributions persist even if $\theta_C = 0$.

If the hadronic currents J_λ are added to the leptonic ones j_λ in (8.10), the effective weak interaction (for momentum-transfer small compared with M_W) becomes

$$-2^{\frac{1}{2}} G_W[(J_\lambda^\dagger + j_\lambda^\dagger)(J^\lambda + j^\lambda) + (J_\lambda^Z + j_\lambda^Z)(J^{Z\lambda} + j^{Z\lambda})] \tag{9.11}$$

(compare (8.39)). Inserting the lepton currents from (2.9) and (8.14), (9.11) produces, among other things, the terms

$$- 2^{\frac{3}{2}} G_{\mathrm{W}} J_{\lambda} [\bar{e} \gamma^{\lambda} \tfrac{1}{2} (1 + \gamma_5) \nu_e + \bar{\mu} \gamma^{\lambda} \tfrac{1}{2} (1 + \gamma_5) \nu_{\mu}] + (\text{Herm. conj.})$$
$$- 2^{\frac{3}{2}} G_{\mathrm{W}} J_{\lambda}^{Z} [\bar{\nu}_e \gamma^{\lambda} \tfrac{1}{2} (1 + \gamma_5) \nu_e + \bar{\nu}_{\mu} \gamma^{\lambda} \tfrac{1}{2} (1 + \gamma_5) \nu_{\mu}]. \quad (9.12)$$

Finally, consider the effective non-leptonic interaction in (9.11). J_{λ} has iso-spin 0, $\frac{1}{2}$ and 1, and J_{λ}^{Z} has iso-spin 0 and 1. Therefore (9.11) has contributions with iso-spin $\frac{3}{2}$ (for any value of θ_{W}).

Observationally, weak non-leptonic transitions with change of iso-spin by $\frac{3}{2}$ are suppressed (Bailin 1971). This cannot be accounted for by the intrinsic structure of the weak interactions in the present model (but see Lee & Treiman 1973, and see §18.6 for possible effects of the strong interactions).

9.3 Comparison with experiment

In the Z-current (9.7), the matrix elements of L_{λ}^{q} and $J_{\lambda}^{\mathrm{em}}$ can be measured in charge-exchange neutrino scattering and electron scattering respectively. But no independent information is available about L_{λ}^{C} in (9.7). To make predictions of the effects of J_{λ}^{Z} this difficulty must somehow be overcome. The following possibilities offer themselves:

(a) If the iso-spin of the hadronic system is known to change (for example, $\nu P \to \nu \Delta^+$), the unknown iso-singlets in (9.7) or (9.9) cannot contribute.

(b) If an average is taken over nucleon charge states (in particular, if the target has zero iso-spin), the interference between iso-spin 0 and 1 in J_{λ}^{Z} is eliminated. Then the known iso-spin 1 contribution provides a lower bound.

(c) For high-energy deep-inelastic inclusive cross-sections, the apparent success of the parton model (Feynman 1972: 259) gives hope that one can reliably calculate the iso-singlet contributions in (9.7). In fact, the simplest version of this model gives zero contribution from the charm and strangeness currents (since there are assumed to be no active λ or c quarks in a nucleon).

(d) It may not be a bad approximation just to neglect the charm current, but there is no rigorous justification of this.

With these general remarks in mind, we now discuss some particular processes. For simplicity, we approximate the Cabibbo angle by zero throughout.

(i) $\bar{\nu}_\mu P \to \bar{\nu}_\mu P$

Although this is the simplest process, there is no way of dealing with the iso-singlet currents in (9.7) save by *neglecting* them (as in (d) above). Then one requires the matrix-elements

$$\langle P, p' | J_\lambda^{\text{em}} | P, p \rangle = \bar{u}' [F_1^{\text{em}}(q^2) \gamma_\lambda - (2m_P)^{-1} F_2^{\text{em}}(q^2) \sigma_{\lambda\nu} q^\nu] u, \tag{9.13}$$

$$\langle \mathcal{N}, p' | L_\lambda^\alpha | \mathcal{N}, p \rangle = \bar{u}' \tfrac{1}{2} \tau^\alpha [F_V \gamma_\lambda + F_A \gamma_\lambda \gamma_5$$
$$- (2m_\mathcal{N})^{-1} F_T \sigma_{\lambda\nu} q^\nu + m_\mathcal{N}^{-1} F_P \gamma_5 q_\lambda] u. \tag{9.14}$$

Here $|P, p\rangle$ is a proton state of momentum p, energy E, and $|\mathcal{N}, p\rangle$ is a nucleon state, $q = p' - p$ and all six form-factors F are functions of q^2. u, u' are Dirac spinors. It is known that, at $q^2 = 0$,

$$\left. \begin{aligned} F_1 &= 1, & F_V &= \tfrac{1}{2}, & F_A &= \tfrac{1}{2} \times 1.25, \\ F_2 &= \mu_P = 1.79, & F_T &= \tfrac{1}{2}(\mu_P - \mu_N) = 1.85. \end{aligned} \right\} \tag{9.15}$$

The term $q_\lambda F_P$ can be neglected, since it gives a contribution proportional to lepton mass-differences. The simplest assumption about the q^2-dependence of the form-factors is that they are all proportional to one-another (which is known to be a good approximation for $F_1^{\text{em}}, F_2^{\text{em}}, F_V, F_T$). Then, in the ratio

$$R_1 = \frac{\sigma(\bar{\nu}_\mu P \to \bar{\nu}_\mu P)}{\sigma(\bar{\nu}_\mu P \to \mu^+ N)}, \tag{9.16}$$

the common q^2-dependence cancels out. For simplicity we go further and neglect all q-dependence. In this approximation R_1 can be read off from (9.7), (9.10), (9.12), (9.13), (9.14) and (9.15):

$$R_1 \simeq \frac{(1 - 4\sin^2\theta_W)^2 + (1.25)^2}{4[1 + (1.25)^2]} \tag{9.17}$$

(the same result comes from (8.40) and (8.42) in the no recoil approximation, provided 1 is replaced by 1.25 in the axial vector term). This rough treatment is good enough to indicate the dependence of R_1 on $\sin^2\theta_W$, and (9.17) is shown in fig. 9.2.

This process is difficult to detect experimentally, since the average proton recoil momentum is low, making the proton hard to detect. A CERN propane bubble-chamber experiment (Cundy *et al.* 1970) gave

$$R_1 = 0.12 \pm 0.06. \tag{9.18}$$

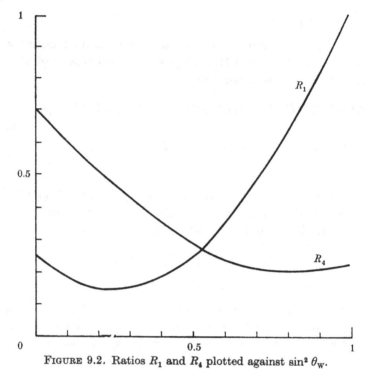

FIGURE 9.2. Ratios R_1 and R_4 plotted against $\sin^2 \theta_W$.

This does not provide much evidence for neutral currents, nor a useful limit on θ_W.

(ii) $\nu \mathcal{N} \to \nu \mathcal{N} \pi$.

These reactions are in some ways more promising than elastic neutrino scattering, for both experimental and theoretical reasons.

First, experimentally, the processes

$$\nu_\mu P \to \nu_\mu N \pi^+, \tag{9.19}$$

$$\nu_\mu P \to \nu_\mu P \pi^0, \tag{9.20}$$

$$\nu_\mu N \to \nu_\mu P \pi^-, \tag{9.21}$$

have been seen in the Argonne National Laboratory 12 foot bubble-chamber filled with liquid hydrogen and deuterium (S. J. Barish *et al.* 1974). A total of 14 candidate events are reported with an estimated background of 4. Results given are

$$R_2 \equiv \frac{\sigma(\nu_\mu P \to \nu_\mu P \pi^0)}{\sigma(\nu_\mu P \to \mu^- P \pi^+)} = 0.51 \pm 0.27, \tag{9.22}$$

$$R_3 \equiv \frac{\sigma(\nu_\mu P \to \nu_\mu N \pi^+)}{\sigma(\nu_\mu P \to \mu^- P \pi^+)} = 0.17 \pm 0.08. \tag{9.23}$$

Previously, limits had been put on R_3 by a CERN propane bubble-chamber experiment (Cundy *et al.* 1970), and on

$$R_4 = \frac{\sigma(\nu_\mu N \to \nu_\mu N \pi^0) + \sigma(\nu_\mu P \to \nu_\mu P \pi^0)}{\sigma(\nu_\mu N \to \mu^- P \pi^0)} \quad (9.24)$$

in a spark-chamber experiment at Brookhaven National Laboratory (W. Lee, 1972). But the interpretation of these results in complex nuclei is bedevilled by the possibility of charge–exchange scattering in the final state (discussed theoretically by Adler, Nussinov and Paschos 1974).

On the theoretical side, Adler (1974) has adapted his earlier successful calculations of photo-production, electro-production and neutrino-production of pions to predict the neutral current pion production. As an example, fig. 9.2 shows Adler's prediction for the ratio R_4 as a function of $\sin^2\theta_W$. Unlike most neutral current effects (compare fig. 8.2 and fig. 9.2) it has a minimum for large $\sin^2\theta_W$, and so it can help to place a lower limit on θ_W. This calculation neglects the final term in (9.8). (This neglect is certainly justified if the iso-spin $\frac{3}{2}$ resonance dominates the final state – see (*a*) at the beginning of this section.)

Actually, in R_4, there is no interference between the iso-spin 0 and 1 parts of the neutral current, because they behave oppositely under charge-symmetry ($P \leftrightarrow N$, $\pi^0 \to -\pi^0$). Therefore, a rigorous lower bound can be set by neglecting all iso-singlet currents in (9.9) (Albright, Lee, Paschos and Wolfenstein 1973); but the bound is rather weak.

Thus the pion production processes already give evidence for neutral currents. They will soon test the Salam–Weinberg model and help to determine θ_W.

(iii) *Inclusive cross-sections*

Here we discuss inelastic neutrino scattering in which the composition of the final hadron state is not determined. In electron scattering, the corresponding inclusive cross-sections seem to exhibit scaling: for large values of the energy- and momentum-transfer the cross-sections are expressible (apart from kinematic factors) in terms of the dimensionless variable $x = (-q^2)/(2m_N \nu)$, where q^2 is the invariant momentum-transfer and ν the energy-transfer in the laboratory frame. The data on neutrino scattering are also consistent with scaling. For example, the total cross-section is proportional to the laboratory neutrino energy.

Conceptually the easiest way to understand scaling is to imagine the lepton to scatter incoherently off 'partons' – constituents which behave as point-like particles (Feynman 1972, Llewellyn Smith 1972). The variable x defined above is interpreted as the proportion of the total nucleon momentum (one must use a frame in which the nucleon has large total momentum) carried by a particular parton. Electron scattering shows that the partons predominately have spin-$\frac{1}{2}$ and no anomalous magnetic moment. In neutrino scattering, the observed ratio

$$\frac{\sigma(\nu_\mu \mathcal{N} \to \mu^- X)}{\sigma(\bar{\nu}_\mu \mathcal{N} \to \mu^+ X)} \simeq 3 \qquad (9.25)$$

(X stands for the final hadron state, whatever it might be), can be understood if the majority of active partons interact in their left-handed state, just like leptons and like the quarks (not anti-quarks) in (9.3). For then the neutrinos interact with partons in the triplet state, and with anti-neutrinos in the singlet state. (A factor of $\frac{1}{3}$ appears in (8.45) and (8.46) from the same cause.)

In more detail, one may assume that the main active partons have the quantum numbers of the quarks. For simplicity, we make the even stronger assumption that the quarks present are only those present in the nucleon according to the quark model (though it should be borne in mind that contributions from λ-quarks or anti-quarks are possible). The mean square charge of the quarks in the proton is then

$$\tfrac{1}{3}[\tfrac{4}{9} + \tfrac{4}{9} + \tfrac{1}{9}] = \tfrac{1}{3},$$

and in the neutron $\tfrac{1}{3}[\tfrac{4}{9} + \tfrac{1}{9} + \tfrac{1}{9}] = \tfrac{2}{9}.$

These numbers seem to accord with electron scattering data. (The p and n quarks, however, account for by no means all the momentum of the nucleon: there must be neutral partons as well.)

With all these assumptions, it is easy to estimate neutral current inclusive cross-sections. Write (9.6) in the quark model, using (9.4) and keeping only p and n quarks:

$$J_\lambda^Z = \bar{p}\gamma_\lambda[(\tfrac{1}{2} - \tfrac{2}{3}\sin^2\theta_W)\tfrac{1}{2}(1 + \gamma_5) - \tfrac{2}{3}\sin^2\theta_W \tfrac{1}{2}(1 - \gamma_5)]\,p$$
$$- \bar{n}\gamma_\lambda[(\tfrac{1}{2} - \tfrac{1}{3}\sin^2\theta_W)\tfrac{1}{2}(1 + \gamma_5) - \tfrac{1}{3}\sin^2\theta_W\tfrac{1}{2}(1 - \gamma_5)]\,n$$
$$+ \text{(contribution from } \lambda \text{ and } c \text{ quarks).} \quad (9.26)$$

In an average over nucleons, p and n quarks occur with equal weight, and left- and right-handed quarks contribute in the ratio 3:1 for neutrino scattering and $1:3$ for anti-neutrino scattering. We want to predict the ratios

$$R_5 = \frac{\sigma(\nu_\mu P \to \nu_\mu X) + \sigma(\nu_\mu N \to \nu_\mu X)}{\sigma(\nu_\mu P \to \mu^- X) + \sigma(\nu_\mu N \to \mu^- X)}, \tag{9.27}$$

$$R_6 = \frac{\sigma(\bar{\nu}_\mu P \to \bar{\nu}_\mu X) + \sigma(\bar{\nu}_\mu N \to \bar{\nu}_\mu X)}{\sigma(\bar{\nu}_\mu P \to \mu^+ X) + \sigma(\bar{\nu}_\mu N \to \mu^+ X)}. \tag{9.28}$$

From (9.26), suitably weighting the quarks, we have (Sehgal 1974)

$$\left.\begin{aligned}
R_5 &= \frac{(\frac{1}{2} - \frac{2}{3}\sin^2\theta_W)^2 + (\frac{1}{2} - \frac{1}{3}\sin^2\theta_W)^2 + \frac{1}{3}[(\frac{2}{3}\sin^2\theta_W)^2 + (\frac{1}{3}\sin^2\theta_W)^2]}{2[(\frac{1}{2})^2 + (\frac{1}{2})^2]}, \\
R_6 &= \frac{\frac{1}{3}[(\frac{1}{2} - \frac{2}{3}\sin^2\theta_W)^2 + (\frac{1}{2} - \frac{1}{3}\sin^2\theta_W)^2] + (\frac{2}{3}\sin^2\theta_W)^2 + (\frac{1}{3}\sin^2\theta_W)^2}{\frac{2}{3}[(\frac{1}{2})^2 + (\frac{1}{2})^2]},
\end{aligned}\right\} \tag{9.29}$$

or
$$\left.\begin{aligned}
R_5 &= \tfrac{1}{2} - \sin^2\theta_W + \tfrac{20}{27}\sin^4\theta_W, \\
R_6 &= \tfrac{1}{2} - \sin^2\theta_W + \tfrac{20}{9}\sin^4\theta_W.
\end{aligned}\right\} \tag{9.30}$$

It is possible to relax the assumption that only p and n quarks are active, and still get useful lower bounds on R_5 and R_6. The nucleon average removes interference between iso-spin 0 and 1, so neglect of the iso-singlet terms in (9.9) gives a lower bound (Paschos and Wolfenstein 1973). The bounds are

$$\left.\begin{aligned}
R_5 &\geqslant \tfrac{1}{2} - \sin^2\theta_W + \tfrac{2}{3}\sin^4\theta_W, \\
R_6 &\geqslant \tfrac{1}{2} - \sin^2\theta_W + 2\sin^4\theta_W.
\end{aligned}\right\} \tag{9.31}$$

The curves of (9.30) and (9.31) are shown in fig. 9.3.

In the actual experiments, there is a cut on the energy-transfer, which means that the factors $\frac{1}{3}$ in (9.29) have to be replaced by slightly bigger factors.

In an experiment using the Gargamelle bubble-chamber (containing freon) at CERN (Hasert et al. 1973b), about two hundred events believed to be neutral current events have been detected. Ratios are quoted
$$\left.\begin{aligned}
R_5 &= 0.22 \pm 0.03, \\
R_6 &= 0.39 \pm 0.05.
\end{aligned}\right\} \tag{9.32}$$

Experiments at the Fermi National Accelerator Laboratory (Benvenuti et al. 1974, Aubert et al. 1974a, Barish et al. 1975) using scintillators interspersed with spark-chambers as target-detectors give similar results, although a rather smaller value of R_6 is quoted. The results are consistent with a value

$$\sin^2\theta_W \simeq 0.3 \quad \text{or} \quad 0.4. \tag{9.33}$$

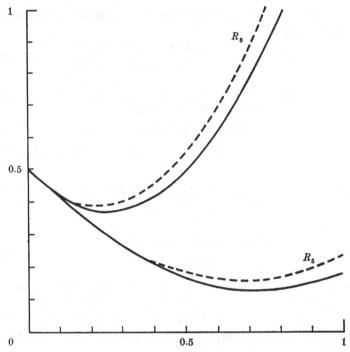

FIGURE 9.3. Lower bounds (continuous lines) and parton model estimates (broken lines) for ratios R_5 and R_6, plotted against $\sin^2 \theta_W$.

The experimental information described in this chapter and the preceding one may be summarized as follows. There is good evidence for neutral weak currents. The Salam–Weinberg model, extended to hadrons as in this chapter, is not contradicted. If it is correct, it is probable that $0.2 < \sin^2 \theta_W < 0.5$.

For some general remarks on neutral currents, see Pais and Treiman (1974).

9.4   Other neutral current effects

Freedman (1975) has pointed out that coherent neutrino–nucleus scattering is possible. At very low energies the matrix-element is, from (9.7), proportional to the nuclear matrix-element of

$$\tfrac{1}{2}T_3 - Q\sin^2\theta_W,$$

which gives an enhancement for heavy nuclei. In astrophysics, this effect may increase the neutrino radiation pressure in supernova formation.

Neutral currents might produce observable parity-violating transitions in heavy atoms (Bouchiat and Bouchiat 1974), and parity-violating effects on the polarization of neutron beams (Stodolsky 1974).

The Z-meson can give measurable contributions to the reaction

$$e^+e^- \to \text{hadrons}$$

at sufficiently high energies (Budny and McDonald 1974).

9.5   The mass of the Higgs particle

The mass of the neutral scalar Higgs particle χ was left completely open in chapter 8. Its interaction with leptons is very weak ((8.28) and (8.32)). The interaction with baryons is probably given by roughly the same type of formula,

$$m_\chi f^{-1} \sim 10^{-3}. \tag{9.34}$$

The exchange of the χ therefore produces very small effects.

Jackiw and Weinberg (1972) have observed that if the χ were very light, say 20 MeV or less in mass, the orbit of the muon in a μ-mesonic atom would lie within the χ-exchange potential. This would cause a relative energy shift of order

$$A m_\mu m_\chi G_W / (Z\alpha) \sim 10^{-4} \tag{9.35}$$

for a nucleus of Z protons and $A - Z$ neutrons.

9.6   Production of charmed states

An inescapable consequence of the Glashow–Iliopoulos–Maiani type of model in this chapter is that charmed particles or states should be produced at sufficiently high energies, either singly by the charged weak current (9.10) or in pairs in, say, e^+e^- annihilation. The lowest lying charmed states cannot be too heavy or the cancellation of fig. 9.1(a) would not be effective in higher-order diagrams. In §16.4 we argue that this sets an upper limit to the mass-difference between the c-quark and the other quarks.

Charmed particles heavier than 2 or 3 GeV would probably decay too quickly to leave a visible track (Snow 1973). In neutrino production, a possible signal is the appearance of two charged leptons, one at the production and one at the decay of the charmed particle. Events with two muons have been reported (Benvenuti et al. 1975). If a charmed particle decays hadronically, the smallness of the Cabibbo angle favours strange particle decay, as can be seen from (9.3). Another possible signal in neutrino production would be an event apparently violating the $\Delta S = \Delta Q$ rule (which is built into the conventional weak current (2.20)). In hadron collisions, the production of a pair of charmed particles might be recognized by a muon from the decay of one and a strange particle from the other.

The present situation in deep-inelastic lepton scattering is that scaling seems to work and that the simple quark-parton model seems to give at least qualitative agreement with the data. (See Aubert *et al* 1974*b* for a possible discrepancy.) If charmed states exist, this success of simple models must be temporary and misleading. At higher energies, where charmed states are produced, a new, true scaling region should be attained; and there the c-quark would have to be included in the quark-parton model. In the simplest model however, in which the active constituents of nucleons are mainly p and n quarks, the new effects would be proportional to θ_C and therefore small.

Consider one specific example. The Adler sum rule (Adler 1966) relates a difference of ν and $\bar{\nu}$ charge-exchange cross-sections to the commutator $[J_0^\dagger, J_0]$. Present data seem to be consistent with the conventional value of that commutator, which is given by $L_\lambda^{'3}$ in (9.2). In the model with charm, the commutator is given instead by J_λ^3 in (9.5). But for small θ_C, the nucleon matrix-elements of J_λ^3 may not be much different from those of $L_\lambda^{'3}$.

Recently, neutral spin-1 narrow resonances have been found at 3.1 and 3.7 GeV (also a wide resonance at about 4.2 GeV). They are probably hadrons (this is suggested by the size of their high-energy photoproduction cross-sections and by the absence of parity-violating asymmetry in their decay), with zero iso-spin and negative G-parity. These objects may be bound-states of a charmed quark and anti-quark, though it is not fully understood why their widths are so small. The reader is referred to volume 34 of *Physical Review Letters*, where a number of experimental and theoretical papers on the new particles are to be found.

10
Feynman's path-integral formulation of quantum mechanics

10.1 Non-relativistic quantum mechanics

We must return to the problem of calculating higher-order effects in weak interactions, since gauge theories claim to make this possible. The quantization of locally gauge-invariant theories is difficult because, for example, the electromagnetic potential $\mathscr{A}_\lambda$ and the Yang–Mills field W_λ^α are gauge-dependent quantities, and do not correspond in a simple way to the genuine dynamical degrees of freedom. The most convenient method of quantizing gauge theories (at least for the purpose of doing covariant perturbation theory) is Feynman's path-integral method (Feynman and Hibbs 1965, Matthews and Salam 1955, Higgs 1956, Polkinghorne 1955). This chapter briefly reviews the path-integral method, largely following the review article of Abers and Lee (1973).

Treat first, for simplicity, a system of one degree of freedom, with a dynamical variable Q, conjugate momentum P (capital latters signify operators here) and Hamiltonian $H(P,Q)$. Denote the Schrödinger and Heisenberg picture by suffices S and H on operators and state-vectors. Then (throughout this chapter we do not use units in which $\hbar = 1$)

$$\left.\begin{array}{l} Q_{\mathrm{H}}(t) = \mathrm{e}^{\mathrm{i}Ht/\hbar} Q_{\mathrm{S}} \mathrm{e}^{-\mathrm{i}Ht/\hbar}, \\ |\alpha t\rangle_{\mathrm{S}} = \mathrm{e}^{-\mathrm{i}Ht/\hbar} |\alpha\rangle_{\mathrm{H}}, \end{array}\right\} \tag{10.1}$$

where α labels a state. In the Schrödinger picture, states are often described by the wave function

$$\langle q | \alpha t \rangle_{\mathrm{S}}, \tag{10.2}$$

where

$$Q_{\mathrm{S}} |q\rangle = q |q\rangle. \tag{10.3}$$

But (10.2) may also be written

$$\langle q t | \alpha \rangle_{\mathrm{H}}, \tag{10.4}$$

where

$$|q t\rangle = \mathrm{e}^{\mathrm{i}Ht/\hbar} |q\rangle \tag{10.5}$$

[75]

and so $\qquad\qquad Q_{\mathrm{H}}(t)|qt\rangle = q|qt\rangle.$ $\qquad\qquad$ (10.6)

The $|qt\rangle$ basis is likened by Dirac to a moving set of axes. The functions

$$\langle q''t''|q't'\rangle \qquad\qquad (10.7)$$

determine the dynamical development of any state:

$$\langle q''|\alpha t''\rangle_{\mathrm{S}} = \int \mathrm{d}q' \langle q''t''|q't'\rangle \langle q'|\alpha t'\rangle_{\mathrm{S}}. \qquad (10.8)$$

We therefore turn our attention to the quantities (10.7).

Split up the time interval (t', t'') into a large number $N+1$ of short intervals τ. Then, by completeness,

$$\langle q''t''|q't'\rangle = \int \prod_{i=1}^{N} \mathrm{d}q_i \langle q''t''|q_N t_N\rangle \langle q_N t_N|q_{N-1}t_{N-1}\rangle \dots \langle q_1 t_1|q't'\rangle.$$
$$(10.9)$$

Using first (10.5), we have

$$\langle q_{i+1} t_{i+1}|q_i t_i\rangle$$
$$= \langle q_{i+1}|e^{-iH\tau/\hbar}|q_i\rangle$$
$$= \langle q_{i+1}|(1-iH\tau/\hbar)|q_i\rangle + O(\tau^2)$$
$$= h^{-1}\int \mathrm{d}p_i\, e^{ip_i(q_{i+1}-q_i)/\hbar}[1 - i\hbar^{-1}\tau H(p_i, \tfrac{1}{2}(q_i+q_{i+1}))] + O(\tau^2),$$
$$(10.10)$$

where the last equality is true at least if $H(P, Q)$ is a function of Q plus a function of P (the p_i-integration provides simply $\delta(q_{i+1}-q_i)$ in the former term and passes to the momentum representation in the latter). Thus finally (10.10) may be written

$$\langle q_{i+1} t_{i+1}|q_i t_i\rangle$$
$$= h^{-1}\int \mathrm{d}p_i \exp\{i\hbar^{-1}[p_i(q_{i+1}-q_i) - \tau H(p_i, \tfrac{1}{2}(q_i+q_{i+1}))]\} + O(\tau^2).$$
$$(10.11)$$

Then (10.9) gives

$$\langle q''t''|q't'\rangle$$
$$= \lim_{N\to\infty} \int \prod_1^N \mathrm{d}q_i \prod_0^N \frac{\mathrm{d}p_i}{h} \exp\left\{i\hbar^{-1}\sum_{i=0}^{N}[p_i(q_{i+1}-q_i) - \tau H(p_i, \tfrac{1}{2}(q_i+q_{i+1}))]\right\},$$
$$(10.12)$$

where $\qquad q_0 = q', \quad q_{N+1} = q'', \quad \tau = (N+1)^{-1}(t''-t').$ $\qquad$ (10.13)

Equation (10.12) may be rewritten formally

$$\langle q''t''|q't'\rangle = \int \mathscr{D}q \, \mathscr{D}\left(\frac{p}{h}\right) \exp i\hbar^{-1}\int_{t'}^{t''} \mathrm{d}t\,[p\dot{q} - H(p,q)], \quad (10.14)$$

where the notation $\int \mathscr{D}q$ represents in some sense an integral over all functions $q(t)$ $(t' \leqslant t \leqslant t'')$. Equation (10.31) becomes

$$q(t') = q', \quad q(t'') = q'', \tag{10.15}$$

but the $\mathscr{D}p$ integration is unrestricted. Note that p, q in (10.14) are classical variables. Problems of operator ordering in $H(P, Q)$ appear to have been lost in (10.14), but they are really concealed in the problem of giving a rigorous definition of the integrals in (10.14).

In the special case
$$H = (2m)^{-1}p^2 + V(q), \tag{10.16}$$

the p_i integrations in (10.12) are easily performed, giving

$$\langle q''t''|q't'\rangle \propto \lim_{N \to \infty} \int \prod_1^N dq_i \exp\left[i\tau\hbar^{-1}\sum_0^N L\left(\frac{q_{i+1}-q_i}{\tau}, \tfrac{1}{2}(q_{i+1}+q_i)\right)\right], \tag{10.17}$$

where the constant of proportionality (formally infinite as $N \to \infty$) is unimportant. (10.17) can be written like (10.14):

$$\langle q''t''|q't'\rangle \propto \int \mathscr{D}q \exp\left[i\hbar^{-1}\int_{t'}^{t''} dt\, L(\dot{q}, q)\right] \tag{10.18}$$

(with (10.15) again applicable). Although (10.14) is more general, (10.18) is adequate for our purposes here.

In field theory applications (10.15) is not the most convenient way of expressing boundary conditions. Instead, it is useful to supplement the Lagrangian by a source function $s(t)$ such that

$$s(t) = 0 \quad \text{for} \quad t > T \quad \text{or} \quad t < -T, \tag{10.19}$$

for some large time interval T. Replace L by

$$L + \hbar s(t)\, q(t). \tag{10.20}$$

Define a functional of $s(t)$

$$Z\{s\} \propto {}_{\mathrm{S}}\langle 0, +\infty|0, -\infty\rangle_{\mathrm{S}}, \tag{10.21}$$

where $|0, t\rangle_{\mathrm{S}}$ is the Schrödinger picture ground state vector for the system defined by (10.20). It turns out that $Z\{s\}$ contains all the information we want.

It can be shown (Higgs 1956, Abers and Lee 1973) that

$$Z\{s\} \propto \lim_{t'' \to \infty\, \mathrm{e}^{-i\delta}} \lim_{t' \to -\infty\, \mathrm{e}^{-i\delta}} \langle q''t''|q't'\rangle, \tag{10.22}$$

where δ is any positive number. The reason is that the time-dependence of $\langle q't'|q''t''\rangle$ for $t' > T$, $t'' < -T$ is governed by terms

$$\exp[-iE_n(t''-T)/\hbar] \quad \text{and} \quad \exp[iE_n(T+t')/\hbar], \qquad (10.23)$$

where E_n is the energy of a stationary state of the system. In the limits in (10.22), only the ground state with minimum energy E_0 survives.

A more convenient way of imposing the boundary condition (10.22) is to add a small negative imaginary part $-\tfrac{1}{2}i\epsilon q^2$ to the Hamiltonian H in (10.14), where ϵ is a small positive number.

Then the exponent

$$-\frac{1}{2}\int_{t'}^{t''} \epsilon q^2 \, dt$$

does duty instead of rotating the time-axis in the complex plane as in (10.22). Therefore we finally define

$$Z\{s\} = \int \mathscr{D}q \exp\left\{i\hbar^{-1}\int_{-\infty}^{\infty} dt[L+\hbar sq+\tfrac{1}{2}i\epsilon q^2]\right\}, \qquad (10.24)$$

and then Z has the interpretation (10.21).

An important property of the path-integral is easily derived. It is

$$\langle q''t''|T(Q_{\mathrm{H}}(t_1)\,Q_{\mathrm{H}}(t_2)\dots Q_{\mathrm{H}}(t_n))|q't'\rangle$$

$$\propto \int \mathscr{D}q\, q(t_1)\dots q(t_n)\exp\left[i\hbar^{-1}\int_{t'}^{t''} L\,dt\right] \quad (t' < t_1, t_2, \dots, t_n < t''),$$

$$(10.25)$$

where $T(\dots)$ denotes a time-ordered product (earlier, further right). The reason that the T-product emerges from the right-hand side of (10.25) is because the $q(t_i)$ have to be inserted into (10.9) in the appropriate places. Only then can the qs be replaced by Qs.

Finally, using (10.24) and (10.25), we get

$$\left[\frac{\delta^n Z}{\delta s(t_1)\dots \delta s(t_n)}\right]_{s=0} \propto i^n \langle 0, +\infty|T(Q_{\mathrm{H}}(t_1)\dots Q_{\mathrm{H}}(t_n))|0, -\infty\rangle.$$

$$(10.26)$$

Here the ground states $\langle 0, \pm\infty|$ have been picked out by the $i\epsilon$ term in (10.24), and we have put $s = 0$ after the functional differentiation so that they are ordinary ground states in the absence of s.

10.2 A scalar field theory

It is easy to generalize §10.1 to include field theories. We state the important equations for the example of a single real scalar field $\phi(x)$.

Equation (10.24) becomes

$$Z\{s(x)\} = \int \mathscr{D}\phi \exp\left\{ i\hbar^{-1} \int d^4x [\mathscr{L} + \hbar s(x)\,\phi(x) + \tfrac{1}{2}i\epsilon\phi^2] \right\}, \quad (10.27)$$

where $\mathscr{L}$ is the Lagrangian density. The ϕ integration is over all functions $\phi(\mathbf{x}, t)$ of space *and* time, because, for each value of $\mathbf{x}$, $\phi(\mathbf{x}, t)$ corresponds to a separate degree of freedom. Boundary conditions (like (10.13)) do not have to be placed on the fields, because of the $i\epsilon$ prescription. The generalization of (10.26) is

$$\left[\frac{\delta^n Z}{\delta s(x_1) \dots \delta s(x_n)} \right]_{s=0} \propto i^n \langle 0, +\infty | T(\phi(x_1) \dots \phi(x_n)) | 0, -\infty \rangle.$$

$$(10.28)$$

It is convenient to define a functional X by

$$Z = \exp(iX/\hbar). \quad (10.29)$$

Then, as we shall immediately confirm, X has the interpretation

$$i\hbar^{-1} \left[\frac{\delta^n X}{\delta s(x_1) \dots \delta s(x_n)} \right]_{s=0} = i^n \langle 0, +\infty | T(\phi(x_1) \dots \phi(x_n)) | 0, -\infty \rangle_{\text{connected}},$$

$$(10.30)$$

where the symbol 'connected' means that only connected Feynman diagrams are to be retained in the perturbation expansion. The unknown constant of proportionality in (10.28) affects X only by an unimportant additive constant.

To be persuaded that (10.30) is right, take as an example $n = 5$, and pick just one of the many terms that result when (10.29) is put into (10.28):

$$\left[\frac{\delta^5 Z}{\delta s(x_1) \dots \delta s(x_5)} \right]_{s=0}$$

$$= i^3 \hbar^{-3} [X]_{s=0} \left[\frac{\delta^2 X}{\delta s(x_1)\,\delta s(x_2)} \right]_{s=0} \left[\frac{\delta^3 X}{\delta s(x_3)\,\delta s(x_4)\,\delta s(x_5)} \right]_{s=0} + \dots. \quad (10.31)$$

If (10.30) is correct, the right-hand side of (10.31) gives graphs made of three disconnected pieces (one of which is a vacuum-to-vacuum graph in this example). Fig. 10.1 gives an example of three such pieces. Thus one sees how the connected diagrams in X build up all diagrams, connected and disconnected, when put into (10.29).

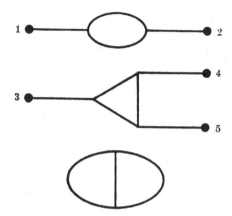

FIGURE 10.1. Examples of a disconnected Feynman diagram.

10.3 Perturbation theory

Suppose the Lagrangian density $\mathscr{L}$ to have the form $\mathscr{L} = \mathscr{L}_0 + \mathscr{L}'$, and $\mathscr{L}'$ is to be used as a perturbation. Write

$$\int \mathscr{L}' d^4 x = S'\{\phi\}.$$

Then (10.27) can be written

$$Z\{s\} = \exp\left[i\hbar^{-1} S'\left\{ -i\frac{\delta}{\delta s} \right\} \right] Z_0\{s\}, \qquad (10.32)$$

where $\qquad Z_0\{s\} = \int \mathscr{D}\phi \exp\left\{ i\hbar^{-1} \int d^4 x[\mathscr{L}_0 + \hbar s\phi + \tfrac{1}{2} i\epsilon\phi^2] \right\}. \qquad (10.33)$

If $\mathscr{L}_0$ is quadratic in ϕ, Z_0 can be explicitly evaluated. Perturbation theory is got by expanding the exponential in (10.32).

For example, let

$$\int d^4 x \mathscr{L}_0 = \frac{1}{2} \int d^4 x (\partial_\lambda \phi \, \partial^\lambda \phi - m^2 \phi^2)$$

$$= -\frac{1}{2} \int d^4 x \, \phi (\Box + m^2) \, \phi. \qquad (10.34)$$

Then the path-integral in (10.33) may be done in analogy with the ordinary multiple integral

$$\int \prod_1^n du_k \exp\{ -\tfrac{1}{2} i \left[u_j(A_{jk} - i\epsilon \delta_{jk}) u_k + 2s_j u_j \right] \}$$
$$= (-i\pi)^{\frac{1}{2}n} (\det A)^{-\frac{1}{2}} \exp[\tfrac{1}{2} i s_j (A - i\epsilon I)^{-1}_{jk} s_k], \qquad (10.35)$$

in which A is a real symmetric matrix and the $i\epsilon$ provides for convergence (this integral may be done by diagonalizing A). Inserting (10.34) into (10.33) and comparing with (10.35), one concludes that

$$Z_0\{s\} \propto \exp\left[\tfrac{1}{2}i\hbar \int d^4x\, d^4y\, s(x)\, (\square + m^2 - i\epsilon)^{-1}\, s(y) \right]. \qquad (10.36)$$

The inverse is found by Fourier transformation:

$$Z_0\{s\} \propto \exp\left[-\tfrac{1}{2}i\hbar \int d^4x\, d^4y\, s(x)\, \Delta_F(x-y)\, s(y) \right], \qquad (10.37)$$

where
$$\Delta_F(x) = (2\pi)^{-4} \int d^4k\, e^{-ik\cdot x}\, (k^2 - m^2 + i\epsilon)^{-1}. \qquad (10.38)$$

The $i\epsilon$ prescription has selected a particular Green's function, the Feynman function Δ_F, as the inverse of the Klein–Gordon operator.

It should now be clear how (10.32), (10.37) and (10.38) together reproduce conventional Feynman perturbation theory, the vertices of the diagrams being determined by the structure of S' in (10.32). For any field, the path-integral in Z_0 generates the inverse of the field's wave-operator. The propagator (2.26) could have been derived in this way. For gauge fields, however, the wave-operator is singular (being nullified by the gauge-increment of the field), and so has no inverse. The above procedure must be modified.

At this point, we establish the useful rule that the contribution to X from a graph with l independent closed loops is proportional to $\hbar^l$. First observe that

$$l = 1 + (\text{number of internal lines}) - (\text{number of internal vertices}).$$

From (10.32) each vertex contributes a factor $\hbar^{-1}$, from (10.37) each line contributes a factor $\hbar$, and (10.29) contributes a single factor $\hbar$.

10.4   One-particle-irreducible functions

The Green's functions generated by $X\{s\}$ have external 'legs' terminating where a source $s(x)$ has been removed by functional differentiation. They are also, in general, one-particle-reducible like the example in fig. 10.2(a), which may be cut in two by severing just one internal line. For many purposes it is more convenient to deal with one-particle-irreducible vertex-functions, which do not have external legs, like the examples in fig. 10.2(b).

Let these be generated by a functional $\Gamma\{\bar\phi\}$ of a classical field $\bar\phi(x)$, just as graphs like fig. 10.2(a) are generated by $X\{s\}$. If $\Gamma\{\bar\phi\}$

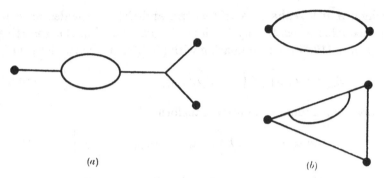

FIGURE 10.2. (a) One-particle-reducible graph.
(b) One-particle-irreducible graphs.

is used as an effective action (we include $S_0\{\tilde\phi\}$ in Γ), and all tree-graphs are drawn using a source term $\int d^4 x s(x)\,\tilde\phi(x)$, then $X\{s\}$ should be generated. We use this statement to define Γ implicitly. Constructing all tree-graphs means solving the 'classical' equations of motion for $\tilde\phi$ (tree-graphs have a factor $\hbar^0$, by the remark at the end of §10.3). Therefore, provided that $\tilde\phi$ is given in terms of s by the classical equations of motion

$$\frac{\delta\Gamma}{\delta\tilde\phi}+\hbar s = 0, \tag{10.39}$$

we have

$$\Gamma\{\tilde\phi\}+\hbar\int d^4 x s(x)\,\tilde\phi(x) = X\{s\}. \tag{10.40}$$

This pair of equations implicitly defines $\Gamma\{\tilde\phi\}$ in terms of $X\{s\}$. (There is an analogy with thermodynamics where, for example, the free-energy $F(T)$ is defined in terms of the internal energy $U(S)$ by

$$F+TS = U.)$$

10.5 S-matrix elements

Since Z, defined by (10.27), generates all Green's functions (10.28), it gives S-matrix elements in particular. These are got by amputating external legs (including self-energy parts on them) from a Green's function, and replacing each leg by a wave-function, satisfying the wave equation (with renormalized mass). This may be expressed by replacing the general source $s(x)$ by a particular form

$$\hat{s}(x) = \lim_{k^2 \to m^2} f(k,x)\,(-\square - m^2), \tag{10.41}$$

where $\qquad (-\Box - m^2) f(k, x) = (k^2 - m^2) f(k, x)$ $\qquad$ (10.42)

and m is the physical, renormalized mass. The Klein–Gordon operator in (10.41) cancels out the pole in the complete propagator corresponding to an external leg. After this has happened, the mass-shell limit, $k^2 \to m^2$, may be taken.

A renormalization factor is contained in the normalization of f. Thus for a spin-0 particle

$$f = Z^{\frac{1}{2}} e^{\pm \mathrm{i} k \cdot x}, \qquad (10.43)$$

where $k_0 > 0$ and the $\pm$ sign corresponds to an ingoing or an outgoing particle respectively.

10.6 Fermions

The path-integral method operates with classical fields only. A fermion field however is quantized with anti-commutation relations, and, in the limit $\hbar \to 0$, such a field anti-commutes with itself (and with its conjugate field and usually with other fermion fields). Thus fermion fields in the Feynman path-integral must be taken to be anti-commuting quantities. The source function, $\sigma(x)$, for a fermion field also must be an anti-commuting quantity and one must remember to use equations like

$$\left. \begin{aligned} \frac{\delta^2}{\delta \sigma(x_1)\, \delta \sigma(x_2)} &= -\frac{\delta^2}{\delta \sigma(x_2)\, \delta \sigma(x_1)}, \\[2mm] \frac{\delta}{\delta \sigma(x)} [\sigma(x_1)\, \sigma(x_2)] &= \delta^4(x - x_1)\, \sigma(x_2) - \delta^4(x - x_2)\, \sigma(x_1). \end{aligned} \right\} \quad (10.44)$$

In practice these rules imply two things. First, initial and final ₋tates are antisymmetric, as is proper for Fermi–Dirac statistics. Second, there is one minus sign for each independent closed fermion loop in a Feynman diagram. This can be verified from the analogue of (10.32), using the rules (10.44). For example, (10.37) becomes

$$\exp\left[-\mathrm{i}\hbar \int \mathrm{d}^4 x\, \mathrm{d}^4 y\, \overline{\sigma}(x)\, S_{\mathrm{F}}(x - y)\, \sigma(y) \right], \qquad (10.45)$$

where $\qquad S_{\mathrm{F}}(x) = (2\pi)^{-4} \int \mathrm{d}^4 k\, e^{-\mathrm{i} k \cdot x} (\gamma \cdot k + m) (k^2 - m^2 + \mathrm{i}\epsilon)^{-1}.$ $\quad$ (10.46)

The third term in the expansion of (10.45) is

$$- (2\hbar^2)^{-1} \int \mathrm{d}^4 x\, \mathrm{d}^4 y\, \mathrm{d}^4 x'\, \mathrm{d}^4 y'\, \overline{\sigma}(x)\, S_{\mathrm{F}}(x - y)\, \sigma(y)$$
$$\times\, \overline{\sigma}(x')\, S_{\mathrm{F}}(x' - y')\, \sigma(y'). \qquad (10.47)$$

In order to provide a fermion closed loop, the fermion analogue of (10.32) must contribute differentials of the form

$$\frac{\delta^2}{\delta\overline{\sigma}_i(z)\,\delta\sigma_j(z)}\frac{\delta^2}{\delta\overline{\sigma}_k(z')\,\delta\sigma_l(z')}. \qquad (10.48)$$

Using the rules (10.44), (10.48) acting on (10.47) yields

$$\hbar^{-2}[S_{\rm F}(z-z')]_{il}\,[S_{\rm F}(z'-z)]_{kj}, \qquad (10.49)$$

with a change of sign (i, j, k, l here are spinor indices).

11

Quantization of gauge fields

11.1 Introduction

The quantization of the Yang–Mills field has been a problem ever since its discovery. Feynman (1963) recognized that the simplest guess for the Feynman rules gives an S-matrix which is not unitary, as explained in §4.2. De Witt (1967, and earlier work referred to therein) and Mandelstam (1968) gave the correct Feynman rules; but Faddeev and Popov (1967), and shortly afterwards Fradkin and Tyutin (1970), showed that the path-integral technique offered the shortest derivation of the rules.

We follow the method of Faddeev and Popov. We restrict ourselves to manifestly covariant gauges. Other gauges depend on an arbitrary 4-vector t_λ. Examples are the Coulomb gauge

$$\partial \cdot W - t \cdot \partial t \cdot W = 0, \quad t^2 = 1$$

and the axial gauge $t \cdot W = 0, \quad t^2 < 0$

($t^2 = 0$ has also been considered as a possibility). The latter has the interesting property that no spurion diagrams (see §11.3) are required. It is, however, a delicate matter to give meaning to the Feynman integrals, because of singular denominators like $(t \cdot k)^{-2}$. These gauges have been discussed by Fradkin and Tyutin (1970), Mohapatra (1971), Tomboulis (1973).

11.2 Separation of the gauge volume

Using the path-integral formalism for, say, the Yang–Mills field, one writes, analogously to (10.27),

$$Z\{s_\lambda^\alpha\} = \int \mathscr{D}W \exp i \int [\mathscr{L} + s_\lambda^\alpha W^{\lambda\alpha}] \, \mathrm{d}^4x \tag{11.1}$$

(we leave out the $i\epsilon$ term, which can be reinstated when required). $\int \mathscr{D}W$ means a path-integration over each of the 4×3 components of W_λ^α (with equal weight). The difficulty in (11.1) is that $\mathscr{L}$, being gauge-invariant, cannot provide a convergence factor in the integration over any one of the 'surfaces' in W_λ^α-space which is generated from a particular value of W by gauge transformations. Since such a 'surface' is infinite, the integral (11.1) diverges. This shows up most simply in the analogue of (10.36). The wave-operator in the Yang–Mills case is $(g_{\lambda\nu}\square - \partial_\nu\partial_\nu)$, and this has no inverse.

To proceed, we must define a notation for path-integration in the space of local gauge transformations. Write a finite gauge transformation as

$$\Omega(x) = \exp\left[\tfrac{1}{2}ig\boldsymbol{\tau}\cdot\boldsymbol{\omega}(x)\right]. \tag{11.2}$$

Let $\mathscr{D}\Omega$ denote an invariant measure of path-integration, so that (for invariant integration, see Gilmore 1974: 78; Hamermesh 1960: 313)

$$\int \mathscr{D}\Omega f\{\Omega\} = \int \mathscr{D}\Omega f\{\Omega\Omega'\} \tag{11.3}$$

for any fixed local gauge transformation Ω'. This is the only property which will be required: it is not necessary to be more explicit about $\mathscr{D}\Omega$.

For infinitesimal gauge transformations, we can take

$$\mathscr{D}\Omega \to \mathscr{D}\omega^\alpha \equiv \mathscr{D}\omega^1\mathscr{D}\omega^2\mathscr{D}\omega^3. \tag{11.4}$$

That this is an invariant volume element can be shown as follows. Let Ω and Ω' and $\Omega'' = \Omega\Omega'$ be given by infinitesimal parameters $\boldsymbol{\omega}$, $\boldsymbol{\omega}'$ and $\boldsymbol{\omega}''$. Then

$$\boldsymbol{\omega}'' = \boldsymbol{\omega} + \boldsymbol{\omega}' - \tfrac{1}{2}g\boldsymbol{\omega} \wedge \boldsymbol{\omega}' \tag{11.5}$$

to first order in $\boldsymbol{\omega}$ and $\boldsymbol{\omega}'$. Since no derivatives are involved, it is sufficient to calculate the Jacobian at any one point of space-time:

$$\det(\partial\omega_\alpha''/\partial\omega_\beta) = \det(\delta_{\alpha\beta} - \tfrac{1}{2}g\epsilon_{\alpha\beta\gamma}\omega_\gamma')$$
$$= 1 + O(\omega'^2). \tag{11.6}$$

This proves the invariance of (11.4).

The integral $\mathscr{D}\Omega$ is, for each x, over a compact region (parametrized, for example, by Euler angles). An integral $\mathscr{D}\omega$ is over an infinitesimal patch; but since it is flat it can be extended to the whole region

$$-\infty < \omega_\alpha < \infty$$

at the cost only of an overall multiplicative factor.

Denote by W_Ω and W_ω the finite and infinitesimal gauge transforms of W. In a similar way to (11.6), one can establish the invariance of the measure in (11.1) under infinitesimal transformations:

$$\int \mathscr{D}Wf\{W\} = \int \mathscr{D}Wf\{W_\omega\}. \tag{11.7}$$

Invariance under infinitesimal transformations is enough to guarantee invariance under finite ones:

$$\int \mathscr{D}W f\{W\} = \int \mathscr{D}W f\{W_\Omega\}. \tag{11.8}$$

We can now return to our main task, which is to smuggle a convergence factor into (11.1). Suppose we could find a functional $\Delta\{W\}$ with the property

$$\int \mathscr{D}\Omega \, \Delta\{W_\Omega\} \exp\left[-\tfrac{1}{2} i\xi^{-1} \int \mathrm{d}^4 x \, (\partial^\lambda (W_\Omega)^\alpha_\lambda)^2 \right] = \text{constant}, \tag{11.9}$$

where ξ is an arbitrary constant. Then, as we show in a moment, the exponential in (11.9) would provide the required convergence, the exponent acting as a gauge-fixing term like (4.10). But first we show how to construct $\Delta\{W\}$.

Define functionals Δ and $\bar{\Delta}$ by

$$[\Delta\{W\}]^{-1} = \int \mathscr{D}\omega \, \delta\{\partial^\lambda (W_\omega)^\alpha_\lambda - B^\alpha\}, \tag{11.10}$$

$$[\bar{\Delta}\{W, B\}]^{-1} = \int \mathscr{D}\Omega \, \delta\{\partial^\lambda (W_\Omega)^\alpha_\lambda - B^\alpha\}, \tag{11.11}$$

where $B^\alpha(x)$ is any field. The symbol $\delta\{\ldots\}$ represents a 'functional Dirac δ-function', that is to say a product of ordinary δ-functions, one for each point of space-time. Δ is independent of B^α because the argument of the δ-function is linear in ω. Δ is, in fact, just a functional determinant. On the other hand, from (11.8), $\bar{\Delta}$ satisfies

$$\bar{\Delta}\{W_\Omega, B\} = \bar{\Delta}\{W, B\} \tag{11.12}$$

for any Ω; but Δ does not obey any similar equation. The two functionals are related by

$$\bar{\Delta}\{W, \partial \cdot W\} = \Delta\{W\}, \tag{11.13}$$

since when $B = \partial \cdot W$ infinitesimal Ω only contribute to (11.11).

Using successively (11.11), (11.12) and (11.13), we deduce

$$1 = \int \mathscr{D}\Omega \, \bar{\Delta}\{W, B\} \, \delta\{\partial \cdot W_\Omega - B\}$$

$$= \int \mathscr{D}\Omega \, \bar{\Delta}\{W_\Omega, B\} \, \delta\{\partial \cdot W_\Omega - B\}$$

$$= \int \mathscr{D}\Omega \, \Delta\{W_\Omega\} \, \delta\{\partial \cdot W_\Omega - B\}. \tag{11.14}$$

The required equation (11.9) follows on inserting (11.14) into the identity

$$\int \mathscr{D}B \exp\left[-\tfrac{1}{2}\mathrm{i}\xi^{-1}\int \mathrm{d}^4x (B^\alpha)^2\right] = \text{constant.} \qquad (11.15)$$

All rigour is absent from these deductions. The ranges of the integrations are not mentioned, nor is the convergence of (11.15).

Having established (11.9), we are ready to return to (11.1). Insert (11.9) under the integral in (11.1) and change the order of integration, to obtain

$$Z\{s\} \propto \int \mathscr{D}\Omega z\{s, \Omega\}, \qquad (11.16)$$

where

$$z\{s, \Omega\} = \int \mathscr{D}W \Delta\{W_\Omega\} \exp\left\{\mathrm{i}\int \mathrm{d}^4x [\mathscr{L} + \mathbf{s}\cdot\mathbf{W} - \tfrac{1}{2}\xi^{-1}(\partial\cdot W_\Omega)^2]\right\}. \qquad (11.17)$$

Making use of the gauge-invariance of $\mathscr{L}$, (11.17) becomes

$$z\{s, \Omega\} = \int \mathscr{D}W \Delta\{W\} \exp\left\{\mathrm{i}\int \mathrm{d}^4x [\mathscr{L} + s_\lambda^\alpha (W_{\Omega^{-1}})^{\lambda\alpha} - \tfrac{1}{2}\xi^{-1}(\partial\cdot W)^2]\right\}. \qquad (11.18)$$

Thus, the only Ω-dependence of z is in the source term.

If z were independent of Ω, we would have succeeded in isolating the divergence of the integral (11.1) in the $\mathscr{D}\Omega$ integration (11.16). Because of the s-dependence of (11.18), this is not so. But suppose for the moment we use $Z\{s\}$ to calculate S-matrix elements only. Then, by (10.41), we replace s in (11.18) by

$$\hat{s}_\lambda^\alpha = \lim_{k^2 \to 0} f_\lambda^\alpha(k, x)\square \qquad (11.19)$$

(or a sum of such terms), where

$$\square f = -k^2 f \qquad (11.20)$$

and

$$\partial^\lambda \hat{s}_\lambda^\alpha = 0. \qquad (11.21)$$

In this case, z *is* independent of Ω. To show this, take first an infinitesimal Ω:

$$\int \mathrm{d}^4x \, \hat{s}_\lambda^\alpha (W_\omega)^{\lambda\alpha} = \int \mathrm{d}^4x \, \hat{\mathbf{s}}_\lambda \cdot [\mathbf{W}^\lambda - g(\boldsymbol{\omega} \wedge \mathbf{W}^\lambda) + \partial^\lambda \boldsymbol{\omega}]. \qquad (11.22)$$

The third term on the right vanishes by (11.21), and the second term vanishes because it gives no pole to cancel the zero coming from (11.19) and (11.20) (such a pole can only arise from a single W-propagator emerging from $\hat{s}$). Thus $\int \mathrm{d}^4x \, \hat{\mathbf{s}}_\lambda \cdot \mathbf{W}^\lambda$ is invariant under infinitesimal gauge transformations, and therefore also under finite ones.

We have shown that

$$z\{\hat{s}, \Omega\} = z\{\hat{s}, 1\} \equiv z\{\hat{s}\}. \tag{11.23}$$

Therefore (11.16) becomes $Z\{\hat{s}\} \propto z\{\hat{s}\}.$ (11.24)

The constant of proportionality here is infinite (a factor of $8\pi^2$ for each point of space-time) but harmless. For S-matrix elements, then, the infinity has been isolated in (11.16).

To define Green's functions in general we *choose* to work with

$$z\{s\} \equiv z\{s, 1\}. \tag{11.25}$$

This means making an arbitrary choice of gauge. We do *not* prove that the Green's functions defined by (11.25) are the same as those coming from $Z\{s\}$. The Green's functions we define are not gauge-independent, and in particular they depend upon the value of ξ in (11.18).

There is another approach (due originally to de Witt) to gauge-field quantization, which seeks to maintain exact gauge-invariance with respect to gauge transformations of a classical background field (Honerkamp 1972, Kluberg–Stern and Zuber 1975b). The gauge-fixing term depends upon the background field. We do not pursue this approach here.

11.3 The Faddeev–Popov function

In order to use (11.18) one must know how to compute the functional Jacobian Δ. A convenient way to do this is to use a representation of the δ-function (generalized, of course, from ordinary δ-functions) in (11.10):

$$[\Delta\{W\}]^{-1} \propto \int \mathscr{D}\eta \, \mathscr{D}\omega \exp\left\{i \int d^4x[-\eta^\alpha\partial_\lambda(W_\omega)^{\lambda\alpha} + \eta^\alpha B^\alpha]\right\}, \tag{11.26}$$

where the η integration runs over all real values of the field $\eta(x)$. As emphasized in the paragraph following (11.6), the ω integration in (11.10) and (11.26) may also be taken over all real values of the field $\omega(x)$. So η and ω are like ordinary fields, and the exponent in (11.26) may be thought of as a sort of 'Lagrangian' for these fields:

$$\mathscr{L}_{\mathrm{S}} = -\boldsymbol{\eta}\cdot[\square\,\boldsymbol{\omega} - g\partial_\lambda(\boldsymbol{\omega} \wedge \mathbf{W}^\lambda) + \partial_\lambda\mathbf{W}^\lambda - \mathbf{B}]. \tag{11.27}$$

We shall call the pair of fields $\boldsymbol{\eta}$, $\boldsymbol{\omega}$ 'spurions'. In the literature they are often called Faddeev–Popov fields or 'ghosts'. The 'Lagrangian' $\mathscr{L}_{\mathrm{S}}$ is unusual in that the first term is not diagonal but contains η

and $\boldsymbol{\omega}$. However, the interaction term converts $\boldsymbol{\omega}$ back into $\boldsymbol{\eta}$, so that there is no problem. The two fields alternate around a closed loop. The last two terms in (11.27), being linear in $\boldsymbol{\eta}$, are irrelevant: they can only begin a spurion line, and there is nowhere for such a line to finish (which would require a term linear in $\boldsymbol{\omega}$). This confirms that Δ is independent of $\mathbf{B}$.

Let the right-hand side of (11.26) generate a functional

$$\exp\left[iY\{W\}\right], \tag{11.28}$$

analogous to (10.29), with Y giving connected graphs only. In (11.18), we actually require

$$\Delta\{W\} = \exp\left[-iY\{W\}\right]. \tag{11.29}$$

Passing from (11.28) to (11.29) is only a matter of placing a minus sign in front of each disconnected $\boldsymbol{\omega}$-$\boldsymbol{\eta}$ closed loop, and presents no difficulty in practice. However, it does mean that $\mathscr{L}_{\mathrm{S}}$ cannot be regarded as a true Lagrangian and incorporated into the exponent of (11.18); for that would give Δ^{-1} not Δ. The power of the path-integral formulation lies here: that it can accommodate under the integral in (11.18) a term like Δ which cannot be put into the exponent as a true Lagrangian.

Nevertheless, the spurion contribution can be regarded as coming from a Lagrangian if $\boldsymbol{\omega}$ and $\boldsymbol{\eta}$ are taken to be anti-commuting fields, like fermions. For then, as explained in §10.6, there automatically results one minus sign for each closed loop. Since $\boldsymbol{\omega}$ and $\boldsymbol{\eta}$ never occur as external lines or represent physical particles, there is no contradiction in having spin-0 fermion fields. In spite of the apparent artificiality of this device, it turns out to be very valuable in formulating Ward identities (§12.2).

Our conclusions may be summarized as follows. The complete Lagrangian density

$$\mathscr{L} + \mathbf{s}_\lambda \cdot \mathbf{W}^\lambda - \tfrac{1}{2}\xi^{-1}(\partial\cdot\mathbf{W})^2 - \boldsymbol{\eta}\cdot[\square\,\boldsymbol{\omega} - g\,\partial^\lambda(\boldsymbol{\omega}\wedge\mathbf{W}_\lambda)], \tag{11.30}$$

in which $\boldsymbol{\eta}$, $\boldsymbol{\omega}$ are regarded as fermion fields, yields the correct S-matrix elements (independent of the choice of ξ). It provides a definition of Green's functions, which are not gauge-invariant (and depend upon ξ).

If the present method is applied to the quantization of electrodynamics, a pair of spurion fields results, but they are not coupled to $\mathscr{A}_\lambda$, so they can be ignored. Thus the use of the gauge-fixing term (3.22), without any other complications, is justified.

11.4   The Feynman rules

The third term in (11.30) is just the gauge-fixing term anticipated in (4.10); and so the W-propagator is that given in (3.24). The $3W$ vertex was written down in (4.14). There is also a $4W$ vertex coming from (4.9) which has the form

$$-g^2[\epsilon_{\alpha\beta\eta}\epsilon_{\gamma\delta\eta}(g_{\lambda\nu}g_{\mu\rho}-g_{\lambda\rho}g_{\mu\nu})+\epsilon_{\alpha\delta\eta}\epsilon_{\beta\gamma\eta}(g_{\lambda\mu}g_{\nu\rho}-g_{\lambda\nu}g_{\mu\rho})$$
$$+\epsilon_{\alpha\gamma\eta}\epsilon_{\delta\beta\eta}(g_{\lambda\rho}g_{\mu\nu}-g_{\lambda\mu}g_{\rho\nu})], \quad (11.31)$$

where $\lambda\alpha$, $\mu\beta$, $\nu\gamma$, $\rho\delta$ are the pairs of Lorentz and isotopic indices.

In Feynman diagrams, we represent a spurion by a dotted line, with an arrow pointing away from the η-end (this arrow has nothing to do with the direction of flow of charge or fermion number). Thus fig. 11.1 (a) represents the propagator

$$\delta_{\alpha\beta}(k^2-i\epsilon)^{-1}. \quad (11.32)$$

The $i\epsilon$ prescription means that we have inserted a convergence factor for the spurion field in the same way as for genuine fields. This is the simplest thing to do, though other prescriptions may be possible.

The W-spurion vertex of fig. 11.1 (b) is found from (11.30) to be (q being ingoing)
$$ig\epsilon_{\alpha\beta\gamma}q_\lambda. \quad (11.33)$$

This expression is not symmetric between the two spurions. The q is the momentum of the line with the outgoing arrow.

Fig. 11.1 (c) is an example of a complete Feynman graph involving a spurion loop.

In §4.2 we saw that, without spurion lines, the Feynman rules were not unitary. 't Hooft (1971 a) shows how to verify explicitly that the spurion lines restore unitarity (as they must according to the formal derivation of §11.3).

11.5   Generalization

The work of the foregoing sections may readily be extended to other gauge theories. It is instructive, and not difficult, to treat the general case. We will simply state the more important results.

Denote the fields collectively as Φ_a, where a labels different fields and different components of each field (whether scalar, spinor or vector). The infinitesimal gauge transformations have the general form

$$(\Phi_\omega)_a = \Phi_a+(I_a^\alpha+g^\alpha T_{ab}^\alpha\Phi_b)\omega^\alpha, \quad (11.34)$$

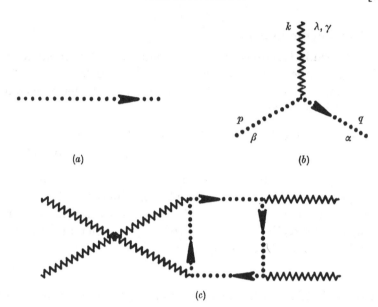

FIGURE 11.1. (a) Spurion line. (b) Spurion-W vertex.
(c) An example.

where I_a^α may be a differential operator. In order for (11.34) to be infinitesimal transformations of a Lie group G, they must satisfy

$$(\Phi_\omega)_{\omega'} - (\Phi_{\omega'})_\omega = \Phi_{\omega''} - \Phi, \qquad (11.35)$$

where

$$\omega_\alpha'' = g_\alpha f_{\alpha\beta\gamma} \omega_\beta \omega_\gamma' \qquad (11.36)$$

(in the notation of §6.5). These equations demand that

$$T_{ab}^\alpha T_{bc}^\beta - T_{ab}^\beta T_{bc}^\alpha = -f^{\alpha\beta\gamma} T_{ac}^\gamma, \qquad (11.37)$$

$$T_{ab}^\alpha I_b^\beta - T_{ab}^\beta I_b^\alpha = -f^{\alpha\beta\gamma} I_a^\gamma. \qquad (11.38)$$

Let the gauge-fixing term, generalizing the third term in (11.30), be

$$\mathscr{L}_G = -\tfrac{1}{2}\xi^{-1}(U_a^\alpha \Phi_a)^2, \qquad (11.39)$$

where U_a^α is also in general a differential operator (terms quadratic in Φ_a might be added to $U_a^\alpha \Phi_a$, but we keep to the simpler form (11.39)).

In terms of these definitions, the spurion Lagrangian density, generalizing the last terms in (11.30), is

$$-\eta^\alpha U_a^\alpha (I_a^\beta + g^\beta T_{ab}^\beta \Phi_b) w^\beta. \qquad (11.40)$$

If G has order N, there are in general N pairs of spurion fields η^α, ω^α (though some of them may be free fields).

These general equations may be applied to the Salam–Weinberg model described in chapter 8. The relevant form of (11.34) is found from (4.21) and (8.22), and (11.40) can be inferred from (8.34) (we will use the first form of (8.34)). The notation

$$
\left.\begin{aligned}
\eta_Z &= \cos\theta_{\mathrm{W}}\,\eta - \sin\theta_{\mathrm{W}}\,\eta_3, \\
\omega_Z &= \cos\theta_{\mathrm{W}}\,\omega - \sin\theta_{\mathrm{W}}\,\omega_3,
\end{aligned}\right\}
\tag{11.41}
$$

helps to simplify the spurion Lagrangian, which comes out to be

$$
\begin{aligned}
&-\eta\,\square\,\omega - \boldsymbol{\eta}\cdot\square\,\boldsymbol{\omega} - \xi M_W^2(\eta_1\omega_1 + \eta_2\omega_2) - \xi M_Z^2\eta_Z\omega_Z \\
&+ g\boldsymbol{\eta}\cdot\partial_\lambda(\boldsymbol{\omega}\wedge \mathbf{W}^\lambda) - \tfrac{1}{2}\xi g M_W\phi_3(\eta_2\omega_1 - \eta_1\omega_2) \\
&- \tfrac{1}{2}\xi g M_Z\eta_Z(\phi_1\omega_2 - \phi_2\omega_1) - \tfrac{1}{2}\xi g M_W\chi(\eta_1\omega_1 + \eta_2\omega_2) \\
&+ \tfrac{1}{2}\xi(g^2 + g'^2)^{\frac{1}{2}} M_Z\chi\eta_Z\omega_Z - \tfrac{1}{2}\xi g M_Z(\eta_1\phi_2 - \eta_2\phi_1)\omega_Z. \quad (11.42)
\end{aligned}
$$

The orthogonal fields to (11.41) are the spurions corresponding to the photon. They are *not* free in (11.42) because of the first interaction term there which contains η_3 and ω_3. The reason the photon has non-trivial spurions is because (8.34) was not chosen to be invariant under electromagnetic gauge transformations. It could be so chosen, but to do so would break the *global SU*(2) × *U*(1) symmetry even when $f = 0$.

12

Ward–Takahashi identities

12.1 Ward–Takahashi identities in gauge theories

Invariance under global transformations has direct physical consequences, like charge conservation or the degeneracy of isotopic multiplets (in the limit of charge independence). Invariance under local transformations (if the group parameters are constrained to tend to zero at infinity in space-time) does not have direct physical consequences. But it does imply extensive relations between Green's functions. In electrodynamics, these are called Ward–Takahashi identities (described in any text-book on quantum electrodynamics). An example of a deduction from these identities is that equality of bare electric charges implies equality of renormalized electric charges.

In non-abelian gauge theories, the Ward–Takahashi identities are rather more complicated, but equally important. They are the necessary and sufficient conditions on the Green's functions, for them to be derived from a locally gauge-invariant Lagrangian (derived, that is, by the process of gauge-fixing described in § 11.3). We use the identities in chapter 14 to show that the process of renormalization is compatible with the local gauge-invariance.

If a Lagrangian is invariant under a global symmetry, the consequence for the one-particle-irreducible vertices is simple: the generating functional Γ (see § 10.4) is invariant under the same transformations applied to the classical fields in Γ. For a local gauge symmetry (particularly if it is non-abelian) the situation is more complex. The gauge-fixing term breaks the invariance, and the spurion 'Lagrangian' is not (in the ordinary sense) gauge-invariant. The generalized Ward–Takahashi identities for non-abelian gauge theories were first formulated in a rather complicated way ('t Hooft 1971a, Slavnov 1972a, Taylor 1971, Lee 1973, 't Hooft and Veltman 1972b). Fortunately the formulation of the identities has been simplified by a device due to Becchi, Rouet and Stora (1974) (see also Kluberg-Stern and Zuber 1975a). This device reduces the effective transformations to global ones, which are, however, non-linear and which mix genuine fields with spurion fields (with anti-commutation relations for the latter).

12.2   The transformations of Becchi, Rouet and Stora

We begin, for simplicity, with a pure Yang–Mills theory. The effective Lagrangian can be written (generalizing (11.30) for an arbitrary group G)

$$\mathscr{L} - \tfrac{1}{2}\xi^{-1}(\partial^\lambda W_\lambda^\alpha)^2 - \eta^\alpha \partial^\lambda D_\lambda^{\alpha\beta} \omega^\beta, \tag{12.1}$$

where $D_\lambda^{\alpha\beta}$ is the appropriate covariant derivative. As explained at the end of §11.3, the spurion term may be regarded as a genuine part of the Lagrangian, provided that η and ω are treated like fermion fields.

The classical Lagrangian $\mathscr{L}$ is invariant under (see (4.4))

$$W_\lambda^\alpha \to W_\lambda^\alpha + \delta W_\lambda^\alpha \equiv W_\lambda^\alpha + D_\lambda^{\alpha\beta} \omega^\beta. \tag{12.2}$$

Suppose we try to identify the group parameters ω in (12.2) with the spurion field ω in (12.1). Then we meet the difficulty that the left-hand side of (12.2) involves a boson field while the right-hand side contains an anti-commuting one. We can get around this by replacing (12.2) by

$$\delta W_\lambda^\alpha \equiv \zeta D_\lambda^{\alpha\beta} \omega^\beta, \tag{12.3}$$

where ζ is a constant anti-commuting quantity (so that $\zeta^2 = 0$), which may conveniently be regarded as infinitesimal.

Under (12.3) the gauge-fixing term in (12.1) transforms by an amount

$$-\xi^{-1}(\partial^\mu W_\mu^\alpha)\,\zeta\,\partial^\lambda D_\lambda^{\alpha\beta}\omega^\beta. \tag{12.4}$$

Invariance is restored if one defines η to have the transformation

$$\delta\eta^\alpha \equiv -\xi^{-1}\zeta\partial^\lambda W_\lambda^\alpha. \tag{12.5}$$

Note that both sides are anti-commuting quantities.

Finally, the third term in (12.1) is not invariant under (12.3) and (12.5) because of the variation (no summation over α)

$$(\delta D_\lambda^{\alpha\beta})\,\omega^\beta = \zeta g^\alpha f^{\alpha\beta\gamma}(D_\lambda\omega)^\beta\,\omega^\gamma$$

$$= \zeta g^\alpha f^{\alpha\beta\gamma}(\partial_\lambda\omega^\beta)\,\omega^\gamma + \zeta g^\alpha f^{\alpha\beta\gamma} g^\beta f^{\beta\delta\epsilon} W_\lambda^\delta \omega^\epsilon\,\omega^\gamma. \tag{12.6}$$

In these equations, care must be taken with the order of the anti-commuting quantities. Using the Jacobi identity (4.20) and anti-commutativity again, (12.6) can be written

$$(\delta D_\lambda^{\alpha\beta})\,\omega^\beta = -\partial_\lambda(\delta\omega^\alpha) - g^\alpha f^{\alpha\beta\gamma} W_\lambda^\beta \delta\omega^\gamma, \tag{12.7}$$

where

$$\delta\omega^\alpha \equiv -\tfrac{1}{2}\zeta g^\alpha f^{\alpha\beta\gamma}\omega^\beta\omega^\gamma. \tag{12.8}$$

If we now regard (12.8) as giving the transformation law for the spurion field ω, then (12.7) states that

$$\delta(D_\lambda \omega)^\alpha = (\delta D_\lambda^{\alpha\beta})\,\omega^\beta + D_\lambda^{\alpha\beta}(\delta\omega)^\beta = 0. \qquad (12.9)$$

Thus (12.1) is invariant under the combined transformations (12.3), (12.5) and (12.8).

From (12.8) and the Jacobi identity, we have

$$\delta(f^{\alpha\beta\gamma}\omega^\beta\omega^\gamma) = -\zeta f^{\alpha\beta\gamma}g^\beta f^{\beta\delta\epsilon}\omega^\gamma\omega^\delta\,\omega^\epsilon = 0 \qquad (12.10)$$

(the anti-commuting property produces the cyclic permutations of the terms in the Jacobi identity).

These transformations sum up all information about the theory in a truly remarkable manner. Equation (12.8) fixes the symmetry group by telling us the structure constants. Equation (12.3) states the transformation law of the field. Equation (12.5) specifies the gauge-fixing term. Equation (12.10) contains the Jacobi identities in a concise form.

We stress yet again that the fermion character of the spurion fields, introduced in § 11.3 as a somewhat artificial trick, has proved crucial. For example, the structure constants are normally defined in terms of two different infinitesimal transformations, yet (12.8) defines them in terms of the single field ω.

The above transformations have a superficial resemblance to the recently discovered supersymmetry transformations (Wess and Zumino 1974), as these too mix boson and fermion fields. However, the fermions in supersymmetry theories are genuine particles with half-integral spin; so the resemblance is probably without significance.

12.3   The generalized Ward–Takahashi identities

Our next aim is to find the conditions implied by the invariance found in § 12.2, on the functional Γ which generates one-particle-irreducible vertices (§ 10.4). Because (12.3) and (12.8) are non-linear, we cannot do this directly. Following Kluberg–Stern and Zuber (1975a), we first introduce extra sources for the non-linear operators in (12.3) and (12.8). Including also sources for each of the fields, we add to (12.1) the terms

$$s_\lambda^\alpha W^{\lambda\alpha} + \omega^\alpha x^\alpha + \eta^\alpha y^\alpha + u_\lambda^\alpha(D^\lambda\omega)^\alpha - \tfrac{1}{2}v^\alpha g^\alpha f^{\alpha\beta\gamma}\omega^\beta\omega^\gamma, \qquad (12.11)$$

in which x, y and u are anti-commuting sources. From (12.9) and (12.10), the coefficients of u and v are each invariant. The generating functional is now a functional of all five sources, and as in (10.29) we define X by

$$z\{s, x, y; u, v\} = \exp iX\{s, x, y; u, v\}. \tag{12.12}$$

(As in (11.18), we use the symbol z not Z because the Lagrangian (12.1) is used instead of $\mathscr{L}$.)

To obtain the identities, make the substitutions

$$W \to W + \delta W, \quad \omega \to \omega + \delta\omega, \quad \eta \to \eta + \delta\eta \tag{12.13}$$

in the path-integral which gives (12.12). The Jacobian of the transformation from the old to the new variables is unity. This follows because

$$\left. \begin{aligned} \frac{\delta[W_\lambda^\alpha(x) + \delta W_\lambda^\alpha(x)]}{\delta W_\nu^\beta(y)} &= \delta^4(x-y)\,\delta_\lambda^\nu\,(\delta^{\alpha\beta} + \zeta g^\gamma f^{\alpha\beta\gamma}\omega^\gamma), \\ \frac{\delta[\omega^\alpha(x) + \delta\omega^\alpha(x)]}{\delta\omega^\beta(y)} &= \delta^4(x-y)\,(\delta^{\alpha\beta} - \zeta g^\gamma f^{\alpha\beta\gamma}\omega^\gamma). \end{aligned} \right\} \tag{12.14}$$

The Jacobian involves the determinants of the matrices on the right, and these are each equal to one (we work to first order in ζ, because $\zeta^2 = 0$). It follows that the functional integral is unchanged by (12.13). Since (12.1) and the u, v terms in (12.11) are invariant, we deduce that

$$\int \mathscr{D}W\mathscr{D}\omega\mathscr{D}\eta \, e^{iS} \int d^4x \, (s_\lambda^\alpha \delta W^{\lambda\alpha} + \delta\omega^\alpha x^\alpha + \delta\eta^\alpha y^\alpha) = 0, \tag{12.15}$$

where S is the complete action integral constructed from (12.1) and (12.11). Using (12.3), (12.5) and (12.8), one can write (12.15) in the form

$$\zeta \int d^4x \left[s_\lambda^\alpha(x) \frac{\partial z}{\delta u_\lambda^\alpha(x)} + \frac{\delta z}{\delta v^\alpha(x)} x^\alpha(x) - \xi^{-1} \left(\partial_\lambda \frac{\delta z}{\delta s_\lambda^\alpha(x)} \right) y^\alpha(x) \right] = 0. \tag{12.16}$$

Since this contains first-order derivatives only (which would not have been the case had we not introduced the sources u and v for the non-linear operators δW and $\delta\omega$), it gives, by (12.12),

$$\int d^4x \left[s_\lambda^\alpha \frac{\delta X}{\delta u_\lambda^\alpha} + \frac{\delta X}{\delta v^\alpha} x^\alpha - \xi^{-1} \left(\partial_\lambda \frac{\delta X}{\delta s_\lambda^\alpha} \right) y^\alpha \right] = 0. \tag{12.17}$$

The next task is to convert (12.17) into a condition on the generating functional Γ. Define Γ (somewhat analogously to (10.40)) by

$$X\{s, x, y; u, v\} = \int d^4x \, (s_\lambda^\alpha \tilde{W}^{\lambda\alpha} + \tilde{\omega}^\alpha x^\alpha + \tilde{\eta}^\alpha y^\alpha) + \Gamma\{\tilde{W}, \tilde{\omega}, \tilde{\eta}; u, v\}. \tag{12.18}$$

Note that the sources u, v have not been transformed. With this definition, (12.17) becomes

$$\int \mathrm{d}^4x \left[\frac{\delta\Gamma}{\delta u^{\lambda\alpha}} \frac{\delta\Gamma}{\delta \tilde{W}_\lambda^\alpha} + \frac{\delta\Gamma}{\delta v^\alpha} \frac{\delta\Gamma}{\delta\tilde{\omega}^\alpha} - \xi^{-1} (\partial^\lambda \tilde{W}_\lambda^\alpha) \frac{\delta\Gamma}{\delta\tilde{\eta}^\alpha} \right] = 0. \qquad (12.19)$$

The last result can be simplified. We use the fact that (12.1) and (12.11) are each linear in η; so that the equation of motion corresponding to η is

$$y^\alpha X - \partial_\lambda \frac{\delta X}{\delta u_\lambda^\alpha} = 0, \qquad (12.20)$$

or

$$\frac{\delta\Gamma}{\delta\tilde{\eta}^\alpha} = -\partial_\lambda \frac{\partial\Gamma}{\delta u_\lambda^\alpha}. \qquad (12.21)$$

Thus (12.19) becomes

$$\int \mathrm{d}^4x \left[\frac{\delta\Gamma}{\delta u^{\lambda\alpha}} \left(\frac{\delta\Gamma}{\delta\tilde{W}_\lambda^\alpha} - \xi^{-1} \partial^\lambda \partial^\nu \tilde{W}_\nu^\alpha \right) + \frac{\delta\Gamma}{\delta v^\alpha} \frac{\delta\Gamma}{\delta\tilde{\omega}^\alpha} \right] = 0. \qquad (12.22)$$

Defining

$$\Gamma = \Gamma' - \tfrac{1}{2}\xi^{-1} \int \mathrm{d}^4x\, (\partial \cdot \tilde{W}^\alpha)^2, \qquad (12.23)$$

(12.22) yields

$$\int \mathrm{d}^4x \left[\frac{\delta\Gamma'}{\delta u^{\lambda\alpha}} \frac{\delta\Gamma'}{\delta\tilde{W}_\lambda^\alpha} + \frac{\delta\Gamma'}{\delta v^\alpha} \frac{\delta\Gamma'}{\delta\tilde{\omega}^\alpha} \right] = 0. \qquad (12.24)$$

The last two equations tell us that the gauge-fixing term in (12.1) appears unchanged in Γ, and the remaining part Γ' obeys the same form of invariance condition as does the remaining part (without the gauge-fixing term) of the Lagrangian (12.1), (12.11).

The identities (12.24) are represented graphically in fig. 12.1(a). The small circle denotes where Γ' has been differentiated with respect to $\tilde{W}$, or with respect to $\tilde{\omega}$. Each arrow on a spurion line (dotted line) represents a gradient (i.e. a momentum 4-vector in momentum space). Note that graphs with the structure of fig. 12.1(b) are not included because Γ is one-particle-irreducible.

The spurion lines never begin within a graph. It follows that fig. 12.1(a) can be decomposed into a hierarchy of identities, according to the number of open spurion lines entering (there may be any number of closed spurion loops). In fig. 12.1(c), one spurion line enters and none leaves. In this case, the second term in (12.24) does not contribute. In fig. 12.1(d), two spurion lines enter and one leaves.

Equation (12.24) is the fundamental identity. It is nonlinear, and the auxiliary functionals $\delta\Gamma'/\delta u$ and $\delta\Gamma'/\delta v$ are not known a priori. They are not entirely independent, however, of the ordinary vertex-

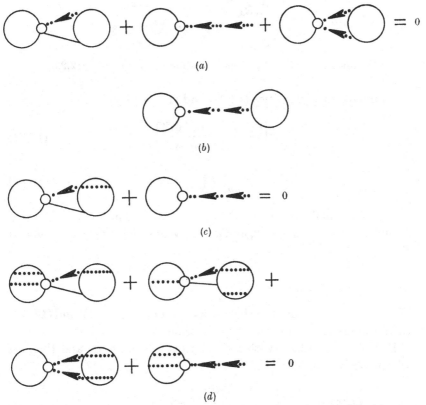

FIGURE 12.1. Graphical representation of Ward identities. (a) The identity
(12.24). (b) A graph which is not included. (c) Case with one open spurion line.
(d) Case with two open spurion lines.

parts of the theory. The connection is provided by (12.21) which, in
terms of (12.23), has the same form:

$$\frac{\delta\Gamma'}{\delta\bar{\eta}^\alpha} + \partial_\lambda \frac{\partial\Gamma'}{\delta u_\lambda^\alpha} = 0. \tag{12.25}$$

This is expressed diagramatically in fig. 12.2, where the first graph
corresponds to u-source vertices contracted with a 4-momentum
(represented by the outgoing arrow), and the small open circle with
the arrow denotes differentiation with respect to $\bar{\eta}$.

12.4 Identities in a general gauge theory

It is easy to adapt the identities of §12.3 to the general gauge theory
of §11.5. The source terms (12.11) become

$$s_a\,\Phi_a + \omega^\alpha x^\alpha + \eta^\alpha y^\alpha + u_a(I_a^\alpha + g^\alpha T_{ab}^\alpha\,\Phi_b)\,\omega^\alpha - \tfrac{1}{2}v^\alpha g^\alpha f^{\alpha\beta\gamma}\,\omega^\beta\,\omega^\gamma. \tag{12.26}$$

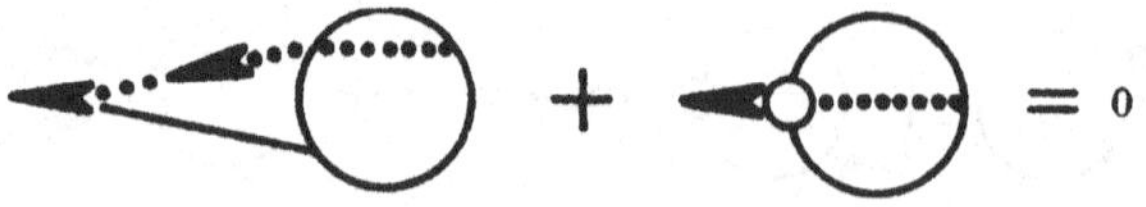

FIGURE 12.2. Graphical representation of the identity (12.25).

The fundamental identities (12.24) and (12.25) read

$$\int d^4x \left[\frac{\delta\Gamma'}{\delta u_a} \frac{\delta\Gamma'}{\delta\tilde{\Phi}_a} + \frac{\delta\Gamma'}{\delta v^\alpha} \frac{\delta\Gamma'}{\delta\tilde{\omega}^\alpha} \right] = 0, \tag{12.27}$$

$$U_a^\alpha \frac{\delta\Gamma'}{\delta u_a} + \frac{\delta\Gamma'}{\delta\tilde{\eta}^\alpha} = 0. \tag{12.28}$$

In the graphs of figs. 12.1 and 12.2, an arrow on a spurion line emerging from a vertex stands for U_a^α, and an arrow entering a vertex stands for I_a^α.

12.5   An example

We give a single example of the use of the identities (12.28) and (12.27): their application to a self-energy function.

Differentiate (12.27) with respect to $\tilde{\Phi}_b(y)$ and $\tilde{\omega}^\beta(z)$ and then set all sources to zero, obtaining

$$\int d^4x \left[\frac{\delta^2\Gamma'}{\delta\tilde{\omega}^\beta(z)\,\delta u_a(x)} \frac{\delta^2\Gamma'}{\delta\tilde{\Phi}_a(x)\,\delta\tilde{\Phi}_b(y)} \right.$$
$$\left. + \frac{\delta^3\Gamma'}{\delta\tilde{\omega}^\beta(z)\,\delta\tilde{\Phi}_b(y)\,\delta u_a(x)} \frac{\delta\Gamma'}{\delta\tilde{\Phi}_a(x)} \right] = 0. \tag{12.29}$$

There are no other terms because spurion lines never terminate; and so $\delta\Gamma'/\delta u$ must be differentiated at least once with respect to $\tilde{\omega}$, and $\delta\Gamma'/\delta v$ at least twice (if the sources are to be put to zero). Similarly (12.28) gives

$$U_a^\alpha \frac{\delta^2\Gamma'}{\delta\tilde{\omega}^\beta(y)\,\delta u_a(x)} + \frac{\delta^2\Gamma'}{\delta\tilde{\omega}^\beta(y)\,\delta\tilde{\eta}^\alpha(x)} = 0, \tag{12.30}$$

where again all sources are put to zero after the differentiation.

The term $\delta\Gamma'/\delta\tilde{\Phi}$ in (12.29) corresponds to the field $\tilde{\Phi}$ disappearing into the vacuum. As will be explained in §14.5, we define the fields in Φ so that this term is zero.

Let us write out the last two equations in more detail, introducing the notation

$$\Phi_a = (W_\lambda^\alpha, \phi_i), \tag{12.31}$$

where the ϕ_i are the Higgs fields (fermions could be included among the ϕ_i if necessary). In this notation

$$U_a^\alpha = (\delta^{\alpha\beta}\,\partial^\lambda, U_i^\alpha), \tag{12.32}$$

since the gauge-fixing term is chosen to have the form

$$-\tfrac{1}{2}\xi^{-1}(\partial \cdot W^\alpha + U_i^\alpha \phi_i)^2. \tag{12.33}$$

Equation (12.29) becomes

$$\int \mathrm{d}^4x\, \frac{\delta^2\Gamma'}{\delta\tilde{\omega}^\beta(z)\,\delta u^{\lambda\alpha}(x)}\, \frac{\delta^2\Gamma'}{\delta\tilde{W}_\lambda^\alpha(x)\,\delta\tilde{W}_\nu^\gamma(y)}$$

$$+ \int \mathrm{d}^4x\, \frac{\delta^2\Gamma'}{\delta\tilde{\omega}^\beta(z)\,\delta u_i(x)}\, \frac{\delta^2\Gamma'}{\delta\tilde{\phi}_i(x)\,\delta\tilde{W}_\nu^\gamma(y)} = 0, \quad (12.34)$$

$$\int \mathrm{d}^4x\, \frac{\delta^2\Gamma'}{\delta\tilde{\omega}^\beta(z)\,\delta u^{\lambda\alpha}(x)}\, \frac{\delta^2\Gamma'}{\delta\tilde{W}_\lambda^\alpha(x)\,\delta\tilde{\phi}_j(y)}$$

$$+ \int \mathrm{d}^4x\, \frac{\delta^2\Gamma'}{\delta\tilde{\omega}^\beta(z)\,\delta u_i(x)}\, \frac{\delta^2\Gamma'}{\delta\tilde{\phi}_i(x)\,\delta\tilde{\phi}_j(y)} = 0, \quad (12.35)$$

and (12.30) becomes

$$\frac{\partial}{\partial x^\lambda}\, \frac{\delta^2\Gamma'}{\delta\tilde{\omega}^\beta(y)\,\delta u_\lambda^\alpha(x)} + U_i^\alpha\, \frac{\delta^2\Gamma'}{\delta\tilde{\omega}^\beta(y)\,\delta u_i(x)} + \frac{\delta^2\Gamma'}{\delta\tilde{\omega}^\beta(y)\,\delta\tilde{\eta}^\alpha(x)} = 0. \tag{12.36}$$

In these equations, $\delta^2\Gamma'/\delta\tilde{W}\,\delta\tilde{W}$, $\delta^2\Gamma'/\delta\tilde{\phi}\,\delta\tilde{\phi}$ and $\delta^2\Gamma'/\delta\tilde{\omega}\,\delta\tilde{\eta}$ are the self-energy functions for, respectively, the vector mesons, the Higgs fields and the spurion; while $\delta^2\Gamma'/\delta\tilde{W}\,\delta\tilde{\phi}$ is a mixing self-energy function. These quantities are related to the auxiliary functions $\delta^2\Gamma'/\delta u\,\delta\tilde{\omega}$. Since

$$\frac{\delta^2\Gamma'}{\delta\tilde{\omega}^\beta(y)\,\delta u_\lambda^\alpha(x)}$$

must have the form $\dfrac{\partial}{\partial x_\lambda}\, F(x-y)\,\delta^{\alpha\beta},$

only the longitudinal parts of the W self-energy and of the W-ϕ mixing function occur in (12.34) and (12.35).

The foregoing equations simplify if there is no spontaneous symmetry-breaking. The quantity $\delta^2\Gamma'/\delta\tilde{\omega}\,\delta u_i$ is zero in this case (at least, if the Lagrangian is invariant under $\phi_i \to -\phi_i$); so that (12.34) just tells us that the W self-energy is transverse.

13

Regularization and anomalies

13.1 The need for regularization

Most perturbation calculations in quantum field theory involve integrals which are ultra-violet divergent. Renormalization can sometimes isolate the divergent parts of the integrals in a few parameters (masses and coupling constants), but nevertheless the divergent integrals have to be manipulated while the convergent parts are being extracted. To do this requires a regularization procedure, that is a way of temporarily modifying the theory so as to make the integrals finite.

The regularized theory must of course violate some physical law, for otherwise it would be a successful finite theory, and no such thing has been found. A good regularization procedure, however, should maintain as many physical properties as possible. For example, the regularization is usually required to be relativistically covariant, so that the final convergent results are guaranteed to be covariant.

In local gauge theories, regularization should maintain the gauge-invariance; for one breaks it at peril of producing results that are not gauge-invariant and so are meaningless. This is particularly clear in Higgs theories, where gauge-invariance is required to assure the equivalence of the renormalizable gauges to the unitary gauge.

Unfortunately, gauge-invariance (particularly under non-abelian gauge groups) defines the theory so closely that it is not easy, by way of regularization, to modify it at all. But one solution to this problem has emerged of great elegance and usefulness ('t Hooft and Veltman 1972a, Bollini and Giambiagi 1972, Ashmore 1972). That solution is dimensional regularization. The dimensions of space-time are temporarily taken to be n where n is a complex number whose real part is less than 4 (sometimes less than 2). Then in, the latter stage of the calculation after renormalization, the limit $n \to 4$ can be taken.

13.2 Rules for dimensional regularization

Theories involving scalar, vector and Dirac fields can be put into n dimensions (initially n is an even integer, later analytic continuation can be done) as follows. Each vector index runs from 0 to $n-1$. The metric tensor is

$$
\begin{aligned}
g_{\lambda\nu} &= 1 && \text{if} \quad \lambda = \nu = 0, \\
&= -1 && \text{if} \quad \lambda = \nu = 1, \dots, (n-1), \\
&= 0 && \text{otherwise,}
\end{aligned}
\tag{13.1}
$$

so that
$$
\delta_\lambda^\lambda = n.
\tag{13.2}
$$

The Dirac γ-matrices satisfy

$$
\gamma_\lambda \gamma_\nu + \gamma_\nu \gamma_\lambda = 2g_{\lambda\nu}
\tag{13.3}
$$

and
$$
\mathrm{Tr}\{I\} = f(n).
\tag{13.4}
$$

All calculations with Dirac matrices and spinors can be reduced to manipulation of matrix products using (13.3) and evaluation of traces using (13.4). Dependence upon n comes from (13.2) and (13.4). In principle one requires $f'(4)$ (a finite contribution from $(n-4)^{-1}f(n)$), but in practice knowledge of this quantity is unimportant, because Ward–Takahashi identities only connect diagrams with equal numbers of fermion loops ('t Hooft and Veltman 1972a). The number $f'(4)$ can be absorbed in renormalization constants.

Local gauge-invariance under an internal symmetry group carries over to n dimensions, so that the Ward–Takahashi identities apply for general n. (γ_5 symmetries will be discussed in §13.4.)

The action

$$
S = \int \mathrm{d}^n x \, \mathscr{L}
\tag{13.5}
$$

is dimensionless (with $\hbar = 1$), and the dimensions of the fields must be adjusted accordingly. Boson fields have dimension

$$
(\text{length})^{1-\frac{1}{2}n}
\tag{13.6}
$$

and fermions
$$
(\text{length})^{\frac{1}{2}-\frac{1}{2}n}.
\tag{13.7}
$$

A coupling constant which would be dimensionless in 4 dimensions ceases to be so in n dimensions. In particular, a constant like the electric charge e or the couplings g and g' in chapter 8 has dimensions

$$
(\text{length})^{\frac{1}{2}n-2}.
\tag{13.8}
$$

It is sometimes convenient to define a dimensionless constant $\hat{g}$ by

$$g = \hat{g}\mu^{2-\frac{1}{2}n}, \tag{13.9}$$

where μ is an arbitrary unit of mass.

If a Feynman integral is treated by the standard method of introducing Feynman parameters, one type of momentum-space integral only is required:

$$\int d^n p [p^2 + 2k \cdot p - K + i\epsilon]^{-r}, \tag{13.10}$$

where k is a fixed 4-momentum, and K is a function of fixed 4-momenta, masses and Feynman parameters. Assuming $n < 2r$, the integral is convergent. Then the origin may be shifted and the contour rotated $90°$ clockwise to obtain the Euclidean integral

$$i \int d^n p' (-p'^2 - K - k^2)^{-r}. \tag{13.11}$$

The surface area of a unit hypersphere in n dimensions is ('t Hooft and Veltman 1972a)
$$2\pi^{\frac{1}{2}n}/\Gamma(\tfrac{1}{2}n), \tag{13.12}$$
so that (13.11) is equal to

$$[i(-1)^r \pi^{\frac{1}{2}n}/\Gamma(\tfrac{1}{2}n)] \int d(p'^2)\, (p'^2)^{\frac{1}{2}n-1}(p'^2 + K + k^2)^{-r}$$
$$= i(-\pi)^{\frac{1}{2}n} [\Gamma(r - \tfrac{1}{2}n)/\Gamma(r)] (-k^2 - K)^{\frac{1}{2}n-r}. \tag{13.13}$$

Other integrals, with p_λ, $p_\lambda p_\nu$ etc. in their numerator can be obtained by differentiation of (13.10) with respect to k.

Expression (13.13) is derived for $\operatorname{Re} n < 2r$. On analytic continuation, it has poles at $n = 2r$, $2r + 2$ etc. The poles at $n = 4$ are the 'divergences' of the integral. If (13.13) is expanded in a Laurent series in $(n - 4)$, the divergent part may be defined to be the $(n - 4)^{-1}$ term, and the convergent part to be the $(n - 4)^0$ term. Thus, for $r = 2$, (13.13) gives

$$-2i\pi^2(n - 4)^{-1} + i\pi^2\{\Gamma'(1) - \ln[\pi(K + k^2)]\} + O(n - 4). \tag{13.14}$$

Renormalization is arranged to cancel out the $(n - 4)^{-1}$ terms, and then the limit $n \to 4$ can be taken.

The definition of the divergent part is arbitrary up to a finite constant; but the above definition is simple and convenient, and we shall use it in the next chapter.

These rules have the rather surprising consequence that

$$\int d^n p (p^2)^{-1} = 0. \tag{13.15}$$

This prescription seems not to lead to any contradiction in the context of dimensional regularization (Capper and Liebrandt 1973, but see also Nouri-Moghadam and Taylor 1975). Strictly speaking, however, there is *no* value of n for which (13.15) is convergent.

Dimensional regularization has also been applied to infra-red divergences (Gastmans and Meuldermans 1973).

13.3 Anomalies

There are two important sorts of symmetry (or approximate symmetry) which dimensional regularization does not respect – that is to say, the Lagrangian has the symmetry for $n = 4$ but no Lagrangian can be found that has it for $n \neq 4$. The first sort consists of γ_5 symmetries and the second of dilatation (or, more generally, conformal) symmetries (associated with the re-scaling of lengths).

It is easy to see why dimensional regularization spoils dilatation invariance. A Lagrangian in 4 dimensions with no masses and with dimensionless coupling constants is formally scale-invariant. But in n dimensions, the constants cease to be dimensionless (see (13.8)), and break the scale-invariance. With γ_5 symmetries, the problem is to find a suitable generalization of γ_5 in n dimensions. We return to this shortly.

Suppose now that a symmetry of one of these types is associated with a Noether current j_λ^a (where a stands for internal or Lorentz indices), which is conserved in 4 dimensions but not in n dimensions. (More generally the divergence in 4 dimensions might be a known operator with finite matrix-elements.) Then

$$\partial^\lambda j_\lambda^a = (n-4) D^a, \qquad (13.16)$$

where D^a is some operator. Taking matrix-elements,

$$iq^\lambda \langle \ \ |j_\lambda^a| \ \ \rangle = (n-4) \langle \ \ |D^a| \ \ \rangle. \qquad (13.17)$$

The problem is that the matrix-element $\langle \ \ |D^a| \ \ \rangle$ may be divergent (calculated in perturbation theory); so that it contains a pole term $(n-4)^{-1}$ like that in (13.14). Then, in the limit $n \to 4$, the right side of (13.17) attains a finite, non-zero value. Thus the expected identity, the vanishing of the left side of (13.17), is not correct. This is the so-called anomaly.

In the case of dilatation invariance, it must be accepted that the apparent invariance of a scale-free Lagrangian is illusory. Indeed this acceptance is at the basis of the Callan–Symanzik equation which controls asymptotic behaviour (see § 18.5). In the same way, some apparent consequences of γ_5 invariances are vitiated by anomalies. If they were not, soft pion theory (§ 5.5) would predict too slow a rate for the decay $\pi^0 \to \gamma\gamma$. (See § 13.4.)

But γ_5 anomalies are disastrous for gauge theories. As we have repeatedly emphasized, the generalized Ward–Takahashi identities are crucial to ensure that the theories make physical sense. Moreover, weak interactions require axial-vector currents; so that γ_5 anomalies generally will be present, spoiling the identities in actual calculations. Let us therefore look at the γ_5 anomalies in more detail, to see what can be done about them.

13.4 γ_5 anomalies

For a good recent review of this subject, see Jackiw (1972).

The simplest example of a Feynman diagram having a γ_5 anomaly is shown in fig. 13.1. It is a triangular fermion loop with two vector vertices and one axial-vector vertex. It is sufficient for our purposes to neglect the fermion mass. Then the integral is

$$S_{\lambda\mu\nu}(p,q) = (2\pi)^{-4} \int d^4k \, \mathrm{Tr} \left[\gamma_5 \gamma_\lambda \gamma \cdot (q+k) \, \gamma_\nu \gamma \cdot k \gamma_\mu \gamma \cdot (k-p) \right]$$
$$\times [(q+k)^2 \, k^2 (k-p)^2]^{-1}. \quad (13.18)$$

For zero mass, both the vector and the axial currents should be divergenceless, and the expected identities are simply

$$(p+q)^\lambda S_{\lambda\mu\nu} = 0, \quad (13.19)$$

$$p^\mu S_{\lambda\mu\nu} = 0, \quad (13.20)$$

$$q^\nu S_{\lambda\mu\nu} = 0. \quad (13.21)$$

We first discuss these by a simple argument which is (apparently) independent of any regularization procedure.

Equation (13.19) can be verified from (13.18):

$$(p+q)^\lambda S_{\lambda\mu\nu} = (2\pi)^{-4} \int d^4k \, \mathrm{Tr} \left[\gamma_5 \gamma_\nu \gamma \cdot k \gamma_\mu \gamma \cdot (k-p) \right] [k^2(k-p)^2]^{-1}$$
$$+ (2\pi)^{-4} \int d^4k \, \mathrm{Tr} \left[\gamma_5 \gamma \cdot (q+k) \gamma_\nu \gamma \cdot k \gamma_\mu \right] [(q+k)^2 \, k^2]^{-1}.$$
$$(13.22)$$

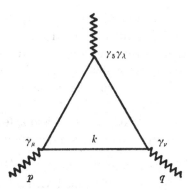

FIGURE 13.1. Fermion triangle with γ_5 anomaly.

Each of these two integrals is a second-rank pseudo-tensor depending on a single 4-vector, and is therefore zero. To verify (13.20) by a similar argument requires a change to a new variable of integration $k' = q + k$, in (13.18). But, because (13.18) is linearly divergent, such a change of variable alters its value by a finite and unambiguous amount. Specifically, a change of variable $k' = k + a$ alters (13.18) by an amount

$$C_{\lambda\mu\nu\sigma} a^\sigma, \tag{13.23}$$

where

$$C_{\lambda\mu\nu\sigma} = (2\pi)^{-4} \int d^4k \frac{\partial}{\partial k^\sigma} \{ \text{Tr}\, [\gamma_5 \gamma_\lambda \gamma \cdot k \gamma_\nu \gamma \cdot k \gamma_\mu \gamma \cdot k] (k^2)^{-3} \}$$

$$= 4(2\pi)^{-4} i \int d^4k \frac{\partial}{\partial k^\sigma} \{ (k^2)^{-2} \epsilon_{\lambda\mu\nu\rho} k^\rho \}$$

$$= -(8\pi^2)^{-1} \epsilon_{\lambda\mu\nu\sigma}, \tag{13.24}$$

where a final factor of i has come from rotating the k_0-contour to get an Euclidean integral and $2\pi^2$ is the volume of the unit sphere in four dimensions (see (13.12)).

We deduce from these considerations that

$$p^\mu S_{\lambda\mu\nu}(p, q) = (8\pi^2)^{-1} \epsilon_{\lambda\mu\nu\sigma} q^\sigma p^\mu, \tag{13.25}$$

and similarly

$$q^\nu S_{\lambda\mu\nu}(p, q) = -(8\pi^2)^{-1} \epsilon_{\lambda\mu\nu\sigma} p^\sigma q^\nu. \tag{13.26}$$

Therefore the tensor

$$T_{\lambda\mu\nu} \equiv S_{\lambda\mu\nu}(p, q) + S_{\lambda\nu\mu}(q, p) + (4\pi^2)^{-1} \epsilon_{\lambda\mu\nu\sigma}(p - q)^\sigma \tag{13.27}$$

satisfies equations like (13.20) and (13.21). Also, (13.27) contains the properly symmetrized form which fig. 13.1 contributes to a Feynman graph. But $T_{\lambda\mu\nu}$ in (13.27) does not obey an equation like (13.19). Instead

$$(p + q)^\lambda T_{\lambda\mu\nu} = (2\pi^2)^{-1} \epsilon_{\lambda\mu\nu\sigma} q^\lambda p^\sigma. \tag{13.28}$$

This is the unambiguous value for the anomaly in the axial identity, assuming that the vector identities are satisfied.

The anomaly cannot be removed by postulating a direct coupling between the three vector-meson fields attached to fig. 13.1. For such a coupling would have the same structure as the last term in (13.27), and this is just the ambiguity we have already allowed for.

In soft pion theory (or PCAC) the pion field is (effectively, at least) proportional to the divergence of the axial current:

$$\pi^\alpha = (f_\pi/m_\pi^2)\,\partial\cdot A^\alpha \quad (\alpha = 1, 2, 3). \tag{13.29}$$

If the axial vertex in fig. 13.1 is identified with A_λ^3, the vector vertices with the electromagnetic current, and the fermions are assumed to be quarks of charge Q, then (13.28) and (13.29) give a contribution to the $\pi^0 \to 2\gamma$ decay amplitude

$$(2\pi^2)^{-1} f_\pi Q^2\, \epsilon_{\lambda\mu\nu\sigma} q^\lambda p^\sigma/m_\pi^2. \tag{13.30}$$

This is in reasonable agreement with the measured lifetime, provided that

$$\Sigma Q^2 \simeq e^2, \tag{13.31}$$

the sum being over all quarks which contribute. If it were not for the anomaly, the π^0 decay amplitude would be predicted, by PCAC and gauge-invariance, to be small. Thus there is, perhaps, some experimental evidence for the reality of the γ_5 anomaly.

So far in this section, we have avoided explicit mention of regularization. It is interesting to see what becomes of dimensional regularization when there are γ_5 vertices. The problem is to define γ_5 in n dimensions. There are two possibilities (Delbourgo and Akyeampong 1974):

$$\gamma_5 \to \gamma_{[\mu_1} \gamma_{\mu_2} \cdots \gamma_{\mu_n]} \quad (n \text{ even}) \tag{13.32}$$

or

$$\gamma_5 \to \gamma_{[\mu_1} \gamma_{\mu_2} \gamma_{\mu_3} \gamma_{\mu_4]}, \tag{13.33}$$

where $[\ldots]$ denotes complete anti-symmetrization. In the first case, the axial-current remains an axial n-vector in n dimensions; but the graph in fig. 13.1 is proportional to

$$\epsilon_{\lambda\mu\nu\mu_4\mu_5\cdots\mu_n} \tag{13.34}$$

and vanishes for $n > 4$ (since there is only one momentum factor to contract with the remaining indices). In this case the idea of analytic continuation is not applicable in any obvious way. In the second case, (13.33), the axial current is a totally antisymmetric tensor of rank $(n-3)$, and its divergence is not zero for $n \neq 4$ (since (13.33) does not

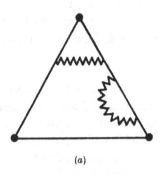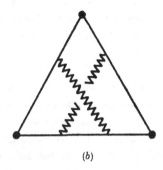

$$(a) \qquad\qquad\qquad\qquad (b)$$

FIGURE 13.2. Higher-order graphs (a) with anomaly,
(b) without anomaly.

anti-commute with the γ-matrices in the Dirac equation). This is exactly the sort of situation envisaged in (13.16) and (13.17), and leads to the expected anomaly (13.28).

In addition to the graph of fig. 13.1, there are anomalies in the AAA triangle graph, $AVVV$ and $AAAV$ square graphs and $AVVVV$, $AAAVV$ and $AAAAA$ pentagon graphs. The square and pentagon anomalies occur only when there are charged currents present. The square anomalies come about because the divergences of these graphs involve linearly divergent integrals. The pentagon anomalies arise because of contact terms introduced to correct the square graphs. (Bardeen 1969, Aviv and Zee 1972, Wess and Zumino 1971).

It is believed that the triangle, square and pentagon fermion loops are the only ones with γ_5 anomalies (Adler and Bardeen 1969). Graphs like fig. 13.2 (a), which contain subdivergences, merely contribute to the renormalization of the parameters in fig. 13.1. Graphs like fig. 13.2 (b) have no anomalies.

This last point may be seen by realizing that, for diagrams with more than one closed loop, there *is* a regularization procedure which respects γ_5 symmetry. In this method of regularization, a fermion Lagrangian, for example $i\bar{\psi}\gamma \cdot D\psi,$

where D is a covariant derivative, is replaced by

$$i\bar{\psi}[1-\Lambda^{-2}D^2]\gamma \cdot D\psi, \qquad\qquad (13.35)$$

Λ being a (large) parameter with the dimensions of mass which eventually tends to infinity (Slavnov 1972b). The propagator behaves like

$$\Lambda^2\gamma \cdot p[p^2(\Lambda^2+p^2)]^{-1} \qquad\qquad (13.36)$$

which makes for convergence. But vertices resulting from the field-dependence of D have factors of Λ^{-2} in them. Bosons are treated similarly. The result is that a graph with l independent loops is proportional to

$$\Lambda^{2l-2}. \tag{13.37}$$

Since these dimensional factors are compensated by momenta in the denominators, this is sufficient to make graphs with $l > 1$ convergent. Thus (13.35) is an adequate regularization procedure for all but single closed loop primitive divergences (a primitive divergence is one containing no divergent sub-integration). The method respects γ_5 invariance, and so it allows no anomalies where it is applicable.

13.5   Gauge theories and γ_5 anomalies

In the presence of γ_5 anomalies properties of gauge theories, which depend for their implementation on the Ward–Takahashi identities, will in general be destroyed. For example, in fig. 13.3, graph (a) does not in general have good high-energy behaviour, and graph (b) does not in general combine unitarity with renormalizability. Plenty of such examples occur in the Salam–Weinberg model (see chapter 8) since it has axial currents in it. Gross and Jackiw (1972) and Korthals Altes and Perrottet (1970) have verified in detail that the expected disasters do indeed occur.

The anomalies are unlikely to be of much practical importance, because they do not occur in low order graphs. But in principle they cannot be tolerated in a gauge theory.

It might be thought that the anomaly (13.28) could be cancelled by the introduction into the Lagrangian of a term of the form

$$C'\phi\epsilon_{\lambda\mu\nu\sigma}\partial^\lambda W^\mu\partial^\sigma W^\nu, \tag{13.38}$$

where ϕ is some spin-0 (Higgs) field. This is quite true, but the snag is that C' has dimensions of length, so that (13.38) is a non-renormalizable interaction.

In any attempt to cancel or avoid the anomalies, it is sufficient to concentrate on fig. 13.1. All other anomalies are related to this one. If an internal symmetry is present with matrices t^α, t^β, t^γ at the three vertices of the figure, the anomaly (13.28) is multiplied by a factor

$$\mathrm{Tr}\left[(t^\alpha t^\beta + t^\beta t^\alpha)\,t^\gamma\right]. \tag{13.39}$$

The simplest way to avoid the anomaly is to choose a group G and a representation for the fermions so that (13.39) is always zero.

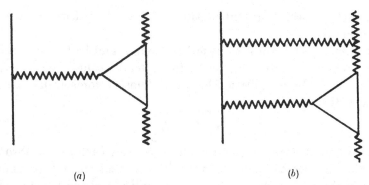

FIGURE 13.3. (a) Graph with bad high-energy behaviour.
(b) Non-unitary or un-renormalizable graph.

Unfortunately, in the Salam–Weinberg $SU(2) \times U(1)$ model, there is a non-zero contribution to (13.39) from

$$t^\alpha = \tfrac{1}{2}\tau^\alpha, \quad t^\beta = \tfrac{1}{2}\tau^\beta, \quad t^\gamma = \tfrac{1}{2}y \qquad (13.40)$$

(that is, γ corresponds to the B vector meson).

If the fermions belong to a real representation, (13.39) must vanish since the ts are then antisymmetric. For example, fermions belonging to an octet representation of $SU(3)$ would have no anomaly. Some groups have only real representations (Georgi and Glashow 1973a). $O(3)$ is an example, but this group cannot accommodate a neutral Z-particle in addition to the photon.

If the anomaly cannot be avoided, perhaps it can be cancelled out. The anomaly is independent of the fermion masses. This opens the way to cancel the lepton triangle with another triangle of heavy (undiscovered) leptons which are right-handed instead of left-handed (so that the axial currents change sign). Call the two sets of lepton fields ψ_L and ψ_R respectively. Then the combinations

$$\psi_\pm = \psi_L \pm \psi_R$$

each have definite parity. Expressed in terms of $\psi_\pm$, it is the mass terms alone which destroy parity conservation. Since mass terms in Higgs type theories partly come from spontaneous symmetry-breaking, parity violation in this model could also be a consequence of the symmetry-breaking (Gross and Jackiw 1972).

The hadrons also would have to be doubled to cancel their contribution to the anomaly. But care must be taken not to identify the π^0 field with the divergence of a complete axial current of the group, or else the successful contribution (13.30) to π^0 decay would be cancelled out.

It is more economical to cancel the lepton and hadron anomalies with one another. In fact, for a simple model containing e^-, ν_e, P and N only, the cancellation happens almost automatically. This is because the doublets

$$\begin{pmatrix} \nu_e \\ e^- \end{pmatrix}, \quad \begin{pmatrix} P \\ N \end{pmatrix} \tag{13.41}$$

have different charge structure, although they each couple through their left-handed components in the charged weak currents (assuming that, in this model, the bare nucleons couple exactly as $V - A$ instead of the observed $V - 1.2A$). To get the charge structures the same, it is simpler to compare the doublets

$$\begin{pmatrix} \nu_e \\ e^- \end{pmatrix}, \quad \begin{pmatrix} \bar{N} \\ -\bar{P} \end{pmatrix}. \tag{13.42}$$

These have opposite handedness, and so it is now clear that they contribute oppositely to the γ_5 anomaly. The same result can be got by observing that the doublets in (13.41) have opposite weak hypercharge y, and so opposite contributions to (13.39) with (13.40).

Can this very attractive idea be extended to include muons and strangeness? In the hadron model described in chapter 9, the four quarks p, n, λ, c are superficially like the four leptons e, ν_e, ν_μ, μ. However, the quark charges in the Gell-Mann–Zweig quark model are in units of $\frac{1}{3}e$ so that cancellation does not take place. One may make the quarks integrally charged only at the cost of all the simplicity of the quark-model classification of hadrons.

There is a possible way forward (but see also §18.4). That is to postulate three quartets of quarks with charges $(1, 0, 0, 1)$, $(1, 0, 0, 1)$, $(0, -1, -1, 0)$. Then the anomalies from two of these quartets cancel among themselves, and the anomaly from the third cancels with the leptons. The zero-charm sector of this model coincides with a variation of the quark model due to Han and Nambu (1965) (see also Nambu and Han 1974).

The point of the Han–Nambu model was partly to resolve a dilemma about quark statistics. The lowest lying baryons belong to the totally symmetric *56* representation of $SU(6)$ (the group combining $SU(3)$ and quark spin). One expects that the quarks should be in s-states. How then can the quarks obey Fermi statistics? In the Han-Nambu model there is an extra degree of freedom: an $SU(3)'$ group which acts on the three multiplets of quarks. It is assumed that the known hadrons are in $SU(3)'$ singlets (an assumption which can be understood if the inter-quark forces are of a suitable character). For baryons, such a singlet is totally antisymmetric, and this is consistent with Fermi statistics.

According to the quark-parton model (see §9.3(iv)), a comparison of the deep-inelastic scattering of electrons and of neutrinos by protons gives an estimate of the sum of the squares of the charges of the quarks in the proton. Present experiments are consistent with the value 1 for this sum, as predicted by the Gell-Mann–Zweig model. In principle, the Han–Nambu model gives a value of 3, but, below the threshold for the production of non-singlet $SU(3)'$ states, it gives the same result as the Gell-Mann–Zweig model (the quarks do not behave completely incoherently below this threshold).

Thus the Han–Nambu model cannot be ruled out. If it is correct, one expects a threshold beyond which $SU(3)'$ non-singlets are produced and qualitative changes in scaling behaviour take place. This threshold is in addition to the charm threshold (§9.6), which is, however, probably harder to observe.

Experimental evidence on the total number of quarks and their charges may come from e^+e^- annihilation experiments. The quark–parton model suggests that the ratio

$$\frac{\sigma(e^-e^- \to \text{hadrons})}{\sigma(e^+e^- \to \mu^+\mu^-)} \to \Sigma Q^2 \tag{13.43}$$

(the sum of the squares of all quark charges) as the energy tends to infinity. This ratio is observed to begin to rise at a centre-of-mass energy of about 3 GeV, and narrow resonances are observed in the same region (see the final paragraph of §9.6). It is possible that the new resonances are $SU(3)'$ non-singlets, but their radiative decay modes seem to be less frequent than expected on this hypothesis.

14

Renormalization of gauge theories

14.1 Renormalization

Renormalization is explained in most text-books of quantum field theory, like Bjorken and Drell (1965) or Bogoliubov and Shirkov (1959). In the context of dimensional regularization, the subject has been reviewed by 't Hooft and Veltman (1972a). We summarize the salient points in this section.

The distinction is made between the renormalized action S^R and the 'bare' action S^B. We will usually express them each in terms of renormalized fields, ϕ^R, but S^R contains renormalized masses and coupling constants $m^R, g^R, \dots$ while S^B contains bare ones $m^B, g^B, \dots$. The renormalized field is defined to have the property that the matrix-element

$$\langle \text{vacuum}|\, \phi\, |\, \text{one-particle state}\rangle$$

is finite. $m^R, g^R, \dots$ are finite but $m^B, g^B, \dots$ are in general infinite (that is, they contain $(n-4)^{-1}$ pole terms in dimensional regularization). Sometimes a bare field

$$\phi^B = Z^{\frac{1}{2}}\phi \tag{14.1}$$

is used, which obeys the canonical equal-time commutation relations. Then Z is in general infinite. The difference between the bare and renormalized action is called the counter-terms, ΔS, so that

$$S^B = S^R + \Delta S. \tag{14.2}$$

The bare action $S^B\{\phi^B\}$ is the true action of the theory. The fundamental result of renormalization theory is that S-matrix elements are given correctly by calculating with $S^R\{\phi^R\}$ but omitting all infinite parts (defined, as in §13.2, to be the pole terms in the expansion about $n = 4$). This means that the counter-terms in ΔS exactly cancel the infinite parts as they arise order by order in perturbation theory.

In some cases it may be convenient to define S^R more carefully. For example, in electrodynamics, one can define the renormalized charge e^R so that the Born approximation for Compton scattering becomes *exact* in the limit of zero momentum for the photons. In this

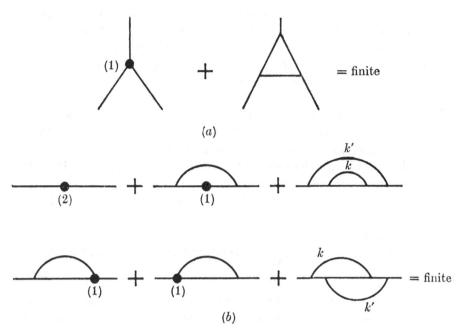

FIGURE 14.1. Renormalization (a) to order $\hbar$, (b) to order $\hbar^2$.

case the counter-term must contain a finite constant in addition to the $(n-4)^{-1}$ pole term. This definition of e^{R} can be used in the Salam–Weinberg model (Ross and Taylor 1973). Normally in what follows, however, we will define the counter-terms to be just the infinite parts (with one exception to be mentioned in §14.5).

The counter-terms in ΔS may be thought of as 2-, 3- and 4-line vertices. The 2-line vertices correspond to mass renormalization and field renormalization (by the factor Z in (14.1)). We represent the counter terms by black blobs in fig. 14.1, where the (l) denotes a term of order $\hbar^l$. (a) and (b) in the figure are examples of renormalization equations. In the third graph of fig. 14.1(b), the k-integration is made finite by the counter-term in the second graph, leaving the k'-integration to be made finite by the $l=2$ counter-term. A like statement cannot be made about the last three graphs. If the k'-integration in the sixth graph is (by arbitrary choice) done first, it is made finite by the fourth. But then to make the k-integration finite, the fifth graph is required as well as the first. This is the phenomenon of overlapping divergences. 't Hooft and Veltman (1972a) explain fairly simply how the process works.

A Lagrangian is renormalizable provided that there is no coupling constant with dimensions (length)^n where $n > 0$, and provided that boson propagators behave like $(k^2)^{-1}$ and fermion propagators like $(k^2)^{-\frac{1}{2}}$ for large k. These conditions ensure that the number of types of divergent integral does not go up indefinitely with increasing orders of perturbation theory.

We now spell out, a little more carefully, how the renormalization process works. Starting from S^R, S^B is determined order-by-order in an expansion in powers of $\hbar$ (that is, by § 10.3, an expansion in terms of the number of independent closed loops in the diagrams). Thus, to order L,

$$g_L^B = \sum_{l=0}^{L} \sum_{m=0}^{l} A_{lm}(g^R) \hbar^l (n-4)^{-m} \tag{14.3}$$

(for shortness, we temporarily use the single symbol g as representative of all the coupling constants and masses). In particular

$$g_0^B = g^R, \quad g_\infty^B = g^B. \tag{14.4}$$

The reason that $m \leqslant l$ in (14.3) is because each integration (one per closed loop) can give rise to at most one power of $(n-4)^{-1}$. At each order, there is a corresponding value of the bare action S_L^B, with

$$S_0^B = S^R, \quad S_\infty^B = S^B. \tag{14.5}$$

The procedure for generating S_{L+1}^B from S_L^B is the following. Let the generating functional Γ of one-particle-irreducible functions (see § 10.4), calculated from S_L^B and including terms up to order $\hbar^{L'}$, be $\Gamma_{L'}(S_L^B)$. In particular

$$\Gamma_0(S_0^B) \equiv \Gamma_0(S^R) = S^R.$$

By hypothesis, $\Gamma_L(S_L^B)$ is finite as $n \to 4$, but $\Gamma_{L+1}(S_L^B)$ contains pole terms. The fundamental result of renormalization theory, which we do not prove here, is that (if S is a renormalizable action) the pole terms in $\Gamma_{L+1}(S_L^B)$ are of a form such that they can be cancelled by counter-terms in (14.2). (This property is not obvious. For instance fig. 14.1(b) might have contributed a term proportional to

$$(n-4)^{-1} \ln (p^2),$$

which cannot be cancelled by any local counter-term.) Assuming the truth of this result, we define

$$S_{L+1}^B = S_L^B - [\Gamma_{L+1}(S_L^B)]_{\mathrm{div}}, \tag{14.6}$$

where the divergent part [ ]$_{\mathrm{div}}$ means the negative powers in a

Laurent series in $(n-4)$ (see §13.2). Then, by construction,

$$\Gamma_{L+1}(S^{B}_{L+1}) = \Gamma_{L+1}(S^{B}_{L}) + S^{B}_{L+1} - S^{B}_{L}$$

is finite.

We close this section by quoting two further special properties of the dimensional regularization version of renormalization.

(i) Divide the renormalized action into a part $S^{R}_{(1)}$ containing dimensionless coupling constants g^{R} only and a part $S^{R}_{(2)}$ containing masses m^{R} (squared masses for bosons) and dimensional coupling constants λ^{R} (between three boson fields):

$$S^{R} = S^{R}_{(1)}(g^{R}) + S^{R}_{(2)}(m^{R}, \lambda^{R}). \tag{14.7}$$

Make a similar decomposition of the bare action:

$$S^{B} = S^{B}_{(1)}(g^{B}) + S^{B}_{(2)}(m^{B}, \lambda^{B}). \tag{14.8}$$

Then each g^{B} depends only upon the g^{R} and upon n and μ^{4-n} (μ was introduced in (13.9)). Also each m^{B} (or $(m^{B})^{2}$) or λ^{B} is a polynomial in the m^{R} and λ^{R}, with coefficients that depend upon the g^{R}, upon n and upon μ^{4-n}. A proof of this result has been given by Collins (1974).

(ii) If $S^{R}_{(1)}$ is invariant under some *global* symmetry group, the same holds for $S^{B}_{(1)}$. To prove this, take the case where $S^{R}_{(2)} = 0$. By (i), this does not alter $S^{B}_{(1)}$, which therefore shares the global symmetry. The converse is also true.

14.2 Renormalization of gauge theories

The renormalization of gauge theories has been discussed by B. W. Lee (1972), Lee and Zinn-Justin (1972, 1973), 't Hooft and Veltman (1972a, b), Zinn-Justin (1975), Kluberg-Stern and Zuber (1975b). The treatment of this section follows Zinn-Justin (1975), where the reformulation of the generalized Ward–Takahashi identities due to Becchi, Rouet and Stora (1975) (see §12.3 and §12.4) makes the task easier.

For finite values of the gauge parameter ξ in (6.17) and (6.19), the gauge theories discussed in previous chapters satisfy the criteria for renormalizability: the coupling constants are dimensionless or have dimensions (length)$^{-1}$, and the propagator (6.19) behaves correctly for large k. The results quoted in §14.1 therefore apply.

One very important result remains to be verified before renormalization is justified for gauge theories. This is that S^{R} in (14.2) can be chosen so that *each of S^{B} and S^{R} is invariant under the local group G.*

The transformation laws may be different in the two cases, but each must constitute a representation of G. This condition severely restricts the independent parameters in ΔS. It has to be shown that the restricted set of counter-terms is sufficient to cancel all the divergent integrals.

Why is this condition necessary? The bare action S^B is *the* complete action, and it is gauge-invariant by definition. But (14.2) is merely a split of S^B into two parts. Why should one part, S^R, by itself be gauge-invariant? The reason is that we do not only require gauge-invariance, we require it order-by-order in perturbation theory. Renormalization is basically a reordering of perturbation theory and we require gauge-invariance order-by-order in an expansion in powers of g. For example, take the tree-approximation in the renormalized expansion. Gauge-invariance is necessary to ensure that the two propagators (6.19) and (6.20) at the pole $k^2 = M^2$ together represent a physical spin-1 particle.

We now begin an inductive proof of the result just stated: if S^R is invariant under a local group G then so also is S_L^B for any L. As shown in §12.3, invariance of an action S implies the generalized Ward–Takahashi identities (12.27) and (12.28). We will first prove that, if S^R is invariant, then S_L^B satisfies these identities. In the next section we explain how the identities imply the invariance of S_L^B.

We concentrate on the first identity (12.27), since (12.28) is linear and therefore simpler. Write (12.27) in the symbolic form

$$\Gamma * \Gamma = 0. \tag{14.9}$$

In order to proceed, we will need to show that S_L^B can be defined so that

$$S_L^B * S_L^B = 0 \tag{14.10}$$

is true *exactly* for each L. This will require the iterative definition (14.6) to be modified by terms $o(\hbar^{L+1})$ – a modification which does not affect the cancellation of divergences or the final value of S^B.

Proceeding inductively, we assume (14.10), which implies, by the argument of §12.3,

$$\Gamma_{L+1}(S_L^B) * \Gamma_{L+1}(S_L^B) = 0.$$

Selecting the divergent terms of order $\hbar^{L+1}$, we obtain

$$S^R * [\Gamma_{L+1}(S_L^B)]_{\mathrm{div}} + [\Gamma_{L+1}(S_L^B)]_{\mathrm{div}} * S^R = 0. \tag{14.11}$$

We must show that (14.11) allows us to define

$$S_{L+1}^B = S_L^B - [\Gamma_{L+1}(S_L^B)]_{\mathrm{div}} + o(\hbar^{L+1})$$

satisfying $\qquad\qquad S^B_{L+1} * S^B_{L+1} = 0 \qquad\qquad$ (14.12)

exactly (it is clearly satisfied with neglect of $o(\hbar^{L+1})$). We indicate how this construction can be done in the case of a single Yang–Mills field (for the general case, see Kluberg-Stern and Zuber 1975b, Dixon 1975).

The first step is the observation that (14.11) has the general solution

$$-[\Gamma_{L+1}(S^B_L)]_{\text{div}} = \sigma_{L+1} S_{\text{cl}} + S^R * F_{L+1} + F_{L+1} * S^R \qquad (14.13)$$

where S_{cl} is the classical Yang–Mills action (the integral of (4.9), without the spurion and source terms) and

$$F_L = \tau_L u^\alpha_\lambda W^{\lambda\alpha} + \tau'_L \omega^\alpha v^\alpha,$$

σ, τ and τ' being divergent constants. That (14.13) satisfies (14.11) can be verified without difficulty. That it is the most general solution can be confirmed by enumerating all possible (dimensionally correct) terms, in the manner of § 14.3. Then, given (14.13), the construction for S^B_{L+1} is

$$S^B_{L+1}(W, \omega, u, v) = S(Y^W_{L+1} W, Y^\omega_{L+1}\omega, Y^u_{L+1} u, Y^v_{L+1} v)$$
$$+ (Y_{L+1} - 1) S_{\text{cl}}(Y^W_{L+1} W, ...)$$

where $\qquad Y_L = 1 + \sum_1^L \sigma_l, \quad Y^W_L = (Y^u_L)^{-1} = 1 + \sum_1^L \tau_l,$

$$(Y^\omega_L)^{-1} = Y^v_L = 1 + \sum_1^L \tau'_l.$$

This construction reproduces (14.6) with neglect of terms $o(\hbar^{L+1})$. It is clearly gauge-invariant (since the re-scaling of the fields does not affect this), and so satisfies (14.12) as required. (For comparison with the notation of § 14.4, note that

$$Y = g^2/g^2_0, \quad Y^W = g^B/g, \quad Y^v = g/(Z_\omega g^B).)$$

14.3   The gauge-invariance of the bare Lagrangian

In the last section, we proved, starting from the action S^R for a gauge theory, that the bare action to any order L, S^B_L, satisfies the generalized Ward–Takahashi identities (12.27) and (12.28). Leaving off the suffix L (or letting $L \to \infty$), we define (in the notation of (11.39)) analogously to (12.23):

$$S^B = S^{B'} - \tfrac{1}{2}\xi^{-1} \int d^4x (U^\alpha_a \Phi_a)^2. \qquad (14.14)$$

Then the identities are

$$\int d^4x \left[\frac{\delta S^{B'}}{\delta u_a} \frac{\delta S^{B'}}{\delta \Phi_a} + \frac{\delta S^{B'}}{\delta v^\alpha} \frac{\delta S^{B'}}{\delta \omega^\alpha} \right] = 0, \tag{14.15}$$

$$U_a^\alpha \frac{\delta S^{B'}}{\delta u_a} + \frac{\delta S^{B'}}{\delta \eta^\alpha} = 0. \tag{14.16}$$

Our next task is to prove that these equations imply S^B to be the complete action for a gauge-theory (that is to prove the converse of the work of §12.3).

In addition to the last two equations, it is helpful to use one further thing. S^R is invariant under the local group G, and, if there were no spontaneous symmetry-breaking, it would also be invariant under the *global* group G (see (ii) of §14.1). This means, if S^R is split up as in (14.7), that $S^R_{(1)}$ is invariant under the global group G, since $S^R_{(1)}$ is unaffected by the spontaneous symmetry-breaking (which yields terms containing the dimensional vacuum-expectation-value F). Therefore, by property (ii) at the end of §14.1, $S^B_{(1)}$ is also invariant under the global group G.

Armed with this fact, and with (14.15) and (14.16), we will now consider first the v-dependence, then the u-dependence, and finally the remainder, of $S^{B'}$. We thus reconstruct S^B to have similar gauge-invariance to S^R.

Dimensionally, the term in S^B containing v cannot contain any derivative or any other field except ω (which is dimensionless). Since spurion lines never terminate (except at u and v), the term must be quadratic in ω as the corresponding term in S^R is (like (12.26)). Therefore this term has the form

$$-\tfrac{1}{2} \int d^4x \, v^\alpha g_\alpha^B Z_\omega^\alpha f_{\alpha\beta\gamma}^B \omega^\beta \omega^\gamma. \tag{14.17}$$

Since (14.17) is part of $S^B_{(1)}$, it is invariant under the global group G, under which v^α and ω^α transform as the regular representation. Since ω^α are anti-commuting quantities,

$$f_{\alpha\beta\gamma}^B = -f_{\alpha\gamma\beta}^B.$$

Differentiating (14.15) with respect to $v^\beta(y)$ and then setting

$$u = v = \Phi = 0,$$

yields the Jacobi identities (12.10) for $f_{\alpha\beta\gamma}^B$. These properties suffice to identify $f_{\alpha\beta\gamma}^B$ with the structure constants of G. By suitably adjusting g_α^B, one obtains

$$f_{\alpha\beta\gamma}^B = f_{\alpha\beta\gamma}, \tag{14.18}$$

where the $f_{\alpha\beta\gamma}$ are the conventionally normalized, totally antisymmetric structure constants.

At this stage, then, (14.17) has established that the ω^α in S^{B} are the parameters of the local group G, and it has defined the bare coupling constants g_α^{B} (which are constant for α within any subgroup of G).

Next, consider the u-dependent term in S^{B} From dimensions and from the fact that spurion lines do not terminate, this term has the form

$$\int \mathrm{d}^4 x \, u_a (I_a^{\mathrm{B}\alpha} + g_\alpha^{\mathrm{B}} T_{ab}^{\mathrm{B}\alpha} \Phi_b) Z_\omega^\alpha \omega^\alpha. \tag{14.19}$$

Differentiating (14.15) with respect to $u_b(y)$, (14.17) and (14.19) are the only terms to contribute, and they give

$$g_\alpha^{\mathrm{B}} T_{ba}^{\mathrm{B}\alpha} \omega^\alpha (I_a^{\mathrm{B}\beta} + g_\beta^{\mathrm{B}} T_{ac}^{\mathrm{B}\beta} \Phi_c) \omega^\beta$$
$$+ \tfrac{1}{2}(I_b^{\mathrm{B}\gamma} + g_\gamma^{\mathrm{B}} T_{bc}^{\mathrm{B}\gamma} \Phi_c) g_\gamma^{\mathrm{B}} f_{\gamma\alpha\beta} \omega^\alpha \omega^\beta = 0 \tag{14.20}$$

(remember that $I_a^{\mathrm{B}\beta}$ contains derivatives operating to the right). Equation (14.20) implies, like (11.37) and (11.38),

$$T_{ab}^{\mathrm{B}\alpha} T_{bc}^{\mathrm{B}\beta} - T_{ab}^{\mathrm{B}\beta} T_{bc}^{\mathrm{B}\alpha} = -f^{\alpha\beta\gamma} T_{ac}^{\mathrm{B}\gamma}, \tag{14.21}$$

$$T_{ab}^{\mathrm{B}\alpha} I_b^{\mathrm{B}\beta} - T_{ab}^{\mathrm{B}\beta} I_b^{\mathrm{B}\alpha} = -f^{\alpha\beta\gamma} I_a^{\mathrm{B}\gamma}. \tag{14.22}$$

It is therefore justified to write (14.19) as

$$\int \mathrm{d}^4 x \, u_a [(\Phi_\omega^{\mathrm{B}})_a - \Phi_a], \tag{14.23}$$

where

$$(\Phi_\omega^{\mathrm{B}})_a = \Phi_a + (I_a^{\mathrm{B}\alpha} + g_\alpha^{\mathrm{B}} T_{ab}^{\mathrm{B}\alpha} \Phi_b) Z_\omega^\alpha \omega^\alpha. \tag{14.24}$$

The second term in (14.19) is part of $S_{(1)}^{\mathrm{B}}$ and so is invariant under the global group G with Φ_a and ω^α transforming in known ways. This together with (14.21), establishes the identification

$$T_{ab}^{\mathrm{B}\alpha} = T_{ab}^\alpha. \tag{14.25}$$

In other words, the $T^{\mathrm{B}\alpha}$ are the matrices of the appropriate representation of G.

Equation (14.16) fixes the spurion part of S^{B} to be

$$-\eta^\alpha U_a^\alpha (I_a^{\mathrm{B}\beta} + g_\beta^{\mathrm{B}} T_{ab}^{\mathrm{B}\beta} \Phi_b) Z_\omega^\alpha \omega^\beta, \tag{14.26}$$

as expected from (14.14) and (14.24).

Finally, putting $\omega = 0$ in (14.15), only the first term survives, and it informs us that the remainder, $S_{\mathrm{cl}}^{\mathrm{B}}$, of $S^{\mathrm{B}'}$ is invariant under (14.19). Thus $S_{\mathrm{cl}}^{\mathrm{B}}$ *is* invariant under the local transformations (14.24), which do constitute *a* representation of the local group G.

We now comment on the differences between S^{R} and S^{B}. In the chain of reasoning we have just been through, arbitrary parameters appear in three places: the g_α^{B} in (14.17), the $I_a^{\mathrm{B}\alpha}$ in (14.19), and in $S_{\mathrm{cl}}^{\mathrm{B}}$ (the couplings and masses of fields other than W_λ^α). Each of these parameters has different values, g_α^{R} etc., in S^{R}.

Further, the normalization of the bilinear terms in S^{R} and S^{B} is different. For example, there are in (14.26) terms proportional to $\eta^\alpha \square \omega^\alpha$. We choose the normalization of the renormalized fields ω^α so that in S^{R} these terms are precisely

$$-\eta^\alpha \square \omega^\alpha. \tag{14.27}$$

Then the renormalized form of (14.19) can be written more explicitly

$$u_\lambda^\alpha \partial^\lambda \omega^\alpha + u_a g_\alpha^{\mathrm{R}} T_{ab}^\alpha \Phi_b \omega^\alpha + u_i I_i^{\mathrm{R}\alpha}\omega^\alpha, \tag{14.28}$$

where i runs over the scalar fields and the $I_i^{\mathrm{R}\alpha}$ have dimension of mass.

Expressed in terms of the renormalized fields, S^{B} contains, instead of (14.28),

$$\sum_\alpha Z_\omega^\alpha [u_\lambda^\alpha \partial^\lambda \omega^\alpha + u_a g_\alpha^{\mathrm{B}} T_{ab}^\alpha \Phi_b \omega^\alpha + u_i I_i^{\mathrm{B}\alpha}\omega^\alpha], \tag{14.29}$$

and, instead of (14.27), $\quad -\sum_\alpha Z_\omega^\alpha \eta^\alpha \square \omega^\alpha. \tag{14.30}$

The Z_ω^α are renormalization constants for the field ω^α, and are independent of α for α within one subgroup of G (by the global invariance of $S_{(1)}$). Similarly, the kinetic term for the field Φ_a in $S_{\mathrm{cl}}^{\mathrm{B}}$ contains a renormalization constant Z_a (independent of a within each irreducible representation of G).

The covariant derivative corresponding to (14.29) is

$$\delta_{ab}\partial_\lambda - \sum_\alpha g_\alpha^{\mathrm{B}} T_{ab}^\alpha W_\lambda^\alpha. \tag{14.31}$$

Defining $\qquad\qquad W_{\lambda\alpha}^{\mathrm{B}} = Z_\alpha^{\frac12} W_{\lambda\alpha}, \tag{14.32}$

(14.31) can be written $\qquad \delta_{ab}\partial_\lambda - g_\alpha^0 T_{ab}^\alpha W_{\lambda\alpha}^{\mathrm{B}}, \tag{14.33}$

where $\qquad\qquad\qquad g_\alpha^{\mathrm{B}} = (Z_\alpha)^{\frac12} g_\alpha^0. \tag{14.34}$

Expression (14.31) or (14.33) defines the way in which g_α^{B} or g_α^0 appears in $S_{\mathrm{cl}}^{\mathrm{B}}$.

Notice that the gauge transformations defined by (14.28) and (14.29) are different, so that $S_{\mathrm{cl}}^{\mathrm{R}}$ and $S_{\mathrm{cl}}^{\mathrm{B}}$ are invariant under *different* transformations, though each set of transformations is a representation of the same local group G. The counter-terms ΔS defined by (14.2) are *not* invariant under any set of local transformations.

FIGURE 14.2. Related spurion graphs.

14.4 Examples of gauge theory renormalization

As a first example of the foregoing general scheme, we take the pure
Yang–Mills theory of chapter 4 (leaving out the fermion field for
simplicity). From now on we drop the suffix R on renormalized
quantities.

The gauge transformation (4.4) contains a single parameter g. The
bare form of (4.4) just has g replaced by g^B. Therefore the bare form of
(4.9) and (11.30) is (adding also the u-source term from (12.11)).

$$-\tfrac{1}{4}Z(\partial_\lambda \mathbf{W}_\nu - \partial_\nu \mathbf{W}_\lambda + g^B \mathbf{W}_\lambda \wedge \mathbf{W}_\nu)^2 - \tfrac{1}{2}\xi^{-1}(\partial\cdot\mathbf{W})^2$$

$$+ Z_\omega(\mathbf{u}_\lambda + \partial_\lambda\eta)\cdot(\partial^\lambda\boldsymbol{\omega} + g^B \mathbf{W}^\lambda \wedge \boldsymbol{\omega}) + \mathbf{s}_\lambda\cdot\mathbf{W}^\lambda. \quad (14.35)$$

The counter-terms proportional to $(Z_\omega - 1)$ cancel (in the lowest order)
the graphs of fig. 14.2, where the first refers to the η term and the second
to the u term. It is clear that the same integral occurs for each graph.

Using the notation

$$Z^{\frac{1}{2}}\,W = W^B, \quad Z_\omega\omega = \omega^B, \quad Z^{-\frac{1}{2}}g^B = g^0, \quad (14.36)$$

(14.35) can alternatively be written

$$-\tfrac{1}{4}(\partial_\lambda \mathbf{W}_\nu^B - \partial_\nu \mathbf{W}_\lambda^B + g^0 \mathbf{W}_\lambda^B \wedge \mathbf{W}_\nu^B)^2 - \tfrac{1}{2}(Z\xi)^{-1}(\partial\cdot\mathbf{W}^B)^2$$

$$+ (\mathbf{u}_\lambda + \partial_\lambda\eta)\cdot(\partial^\lambda\boldsymbol{\omega}^B + g^0 \mathbf{W}^{B\lambda} \wedge \boldsymbol{\omega}^B) + Z^{-\frac{1}{2}}\mathbf{s}^\lambda\cdot\mathbf{W}_\lambda^B. \quad (14.37)$$

The W-propagator obtained from (14.35) is

$$Z^{-1}[-g_{\lambda\nu} + (1 - \xi Z)\,k_\lambda k_\nu/k^2]\,(k^2 + i\epsilon)^{-1}. \quad (14.38)$$

If ξ is chosen to give the renormalized propagator (3.24) a special
form (e.g. $\xi = 1$), (14.38) will not in general share that special form.
An exception is the case $\xi = 0$, which gives a transverse structure to
both renormalized and bare propagators.

As a second example, we take the Salam–Weinberg model of
leptons (chapter 8). Following the general procedure of §14.3, (14.17)
in this case defines two bare coupling constants, g^B and g'^B. Next,
(14.19) defines the bare value, F^B, of the quantity F appearing in
(8.20). It was assumed in chapter 8 that F did not break the $U(1)$
gauge group of electromagnetism, nor the discrete charge-conjugation

symmetry. Therefore the same is true of F^{B}, and F^{B} has the form

$$F^{\mathrm{B}} = 2^{-\frac{1}{2}}\begin{pmatrix} 0 \\ f^{\mathrm{B}} \end{pmatrix} \qquad (14.39)$$

like (8.20).

Finally, one must renormalize those parameters which are not determined by the gauge transformations. Thus m_{e}, m_{μ}, λ in (8.28) and (8.29) are replaced by $m_{\mathrm{e}}^{\mathrm{B}}$, m_{μ}^{B}, λ^{B} in the bare Lagrangian. In addition, renormalization constants $Z_{\mathbf{W}}$, Z_B, Z_{ϕ}, Z_{L}, Z_{R}, Z_{ω}, Z_{ω} multiply the free Lagrangians for the $\mathbf{W}$, B, ϕ, l_{e} (and l_{μ}), e_{R} (and μ_{R}), $\boldsymbol{\omega}$, ω fields respectively (because of parity violation, $Z_{\mathrm{L}} \neq Z_{\mathrm{R}}$; but, by property (ii) at the end of § 14.1, electron and muon renormalizations are equal).

As an example, the complete electron Lagrangian (8.26) and (8.28) has the bare form

$$iZ_{\mathrm{L}}\bar{l}_{\mathrm{e}}\gamma^{\lambda}(\partial_{\lambda} - \tfrac{1}{2}ig^{\mathrm{B}}\boldsymbol{\tau}\cdot\mathbf{W}_{\lambda} + \tfrac{1}{2}ig'^{\mathrm{B}}B_{\lambda})l_{\mathrm{e}} + iZ_{\mathrm{R}}\bar{e}_{\mathrm{R}}\gamma^{\lambda}(\partial_{\lambda} + ig'^{\mathrm{B}}B_{\lambda})e_{\mathrm{R}}$$
$$- 2^{\frac{1}{2}}m_{\mathrm{e}}^{\mathrm{B}}(f^{\mathrm{B}})^{-1}Z_{\mathrm{L}}^{\frac{1}{2}}Z_{\mathrm{R}}^{\frac{1}{2}}[\bar{e}_{\mathrm{R}}(\phi^{\dagger}l_{\mathrm{e}}) + (\bar{l}_{\mathrm{e}}\phi)e_{\mathrm{R}}]. \qquad (14.40)$$

The vector-meson masses in S^{B} come out to be

$$M_W^{\mathrm{B}} = \tfrac{1}{2}g^{\mathrm{B}}f^{\mathrm{B}}, \quad M_Z^{\mathrm{B}} = \tfrac{1}{2}f^{\mathrm{B}}[(g^{\mathrm{B}})^2 + (g'^{\mathrm{B}})^2]^{\frac{1}{2}}, \qquad (14.41)$$

preserving the form of (8.30). One may define a bare Weinberg angle by

$$\tan\theta_{\mathrm{W}}^{\mathrm{B}} = g'^{\mathrm{B}}/g^{\mathrm{B}}.$$

The above prescriptions give the counter-terms ΔS in terms of the fields W_3^{λ} and B^{λ}. It is, of course, possible to re-express them in terms of the physical fields Z^{λ}, A^{λ} using (8.3). For more details, see Ross and Taylor (1973).

Note that the gauge-fixing term (8.34) was designed to give simple renormalized propagators. The same gauge-fixing term appears in S^{B}, but there does not give simple propagators – W-ϕ and B-ϕ mixing takes place in general (of course, this is no inconveience as it is the renormalized propagator that is used in practical calculations). The Landau gauge, $\xi = 0$ in (8.34), is exceptional in being unaffected by renormalization.

14.5   Tadpole renormalization

In this and the next section, we return to a question that was left un-answered in chapter 6: is it correct that the vacuum–expectation-value F of the Higgs field ϕ occurs at the minimum of the potential $V(\phi)$?

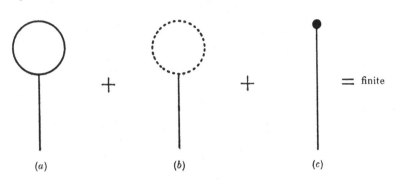

FIGURE 14.3. Tadpole renormalization.

In chapters 6 and 8, we provisionally assumed that it was correct. However this may be for the renormalized Lagrangian, there is certainly nothing in the prescription given in §14.4 for constructing the bare Lagrangian to make the same thing true of that. Taking the Weinberg–Salam model as an example, the bare form of (8.29) has the more general structure

$$\tfrac{1}{4}\lambda^{B}(\phi_1'^{2} + \phi_2'^{2} + \phi_3'^{2} + \chi^{2} + 2f^{B}\chi + f^{B2} - h^{B2})^{2}, \qquad (14.42)$$

where the appearance of the f^{B} factors in the invariant is dictated by the bare transformation properties of $(\phi_1', \phi_2', \phi_3', \chi)$. There is no reason why f^{B} and h^{B} should be equal.

Expression (14.42) contains, among other things, a linear term in χ:
$$\lambda^{B}f^{B}(f^{B2} - h^{B2})\chi, \qquad (14.43)$$

which is indeed required to cancel the divergent 'tadpole' graphs of the sort shown in fig. 14.3(a) and (b). In the examples in the figure, (a) might be a W, Z, ϕ, χ, e or μ loop and (b) is a spurion loop. Graph (c) represents the counter-term from (14.43), so that the sum of all three types of contribution is finite.

We have defined the counter-terms to cancel the infinite parts only, and the result of §14.3 depended upon that fact. Therefore we cannot arrange the sum of the contributions in fig. 14.3 to be zero, only to be finite. This leads to complications. To any graph, 'tadpole tree-graphs' like that in fig. 14.4 can be added, and there are an infinite number of such trees. It is convenient to sum the contribution from these trees exactly. We explain the general method for doing this in the next section.

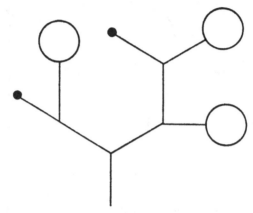

FIGURE 14.4. A tadpole tree.

14.6   The effective potential

Take the generating-functional $\Gamma\{\tilde{\Phi}\}$ of one-particle-irreducible vertices, as defined in §10.4. Pick the part which depends only upon the Higgs multiplet $\phi(x)$ (we drop the tilde off $\tilde{\phi}$). Further, restrict $\phi(x)$ to be a constant field ϕ. Then we obtain from Γ

$$- \int \mathrm{d}^4 x\, U(\phi), \qquad (14.44)$$

where $U(\phi)$ is a function, not a functional any more. The potential V in the renormalized Lagrangian is contained in U:

$$U(\phi; g^{\mathrm{R}}, \dots) = V(\phi; g^{\mathrm{R}}, \dots) + O(\hbar), \qquad (14.45)$$

where $g^{\mathrm{R}}, \dots$ stands for all renormalized coupling constants and masses.

The sum of all tadpole graphs (like fig. 14.3) is just the linear term in $U(\phi)$. As in (6.29), we write

$$\phi = F + \phi', \qquad (14.46)$$

but define F by

$$\left[\frac{\partial U}{\partial \phi}\right]_{\phi = F} = 0, \qquad (14.47)$$

instead of by

$$\left[\frac{\partial V}{\partial \phi}\right]_{\phi = F} = 0. \qquad (14.48)$$

Then, in terms of the field ϕ', there are no tadpoles and so no tadpole-trees. Thus ϕ' is a convenient field to use, and has exactly zero vacuum-expectation-value.

Let us examine the implications for the Salam–Weinberg model. Since (14.48) is not correct, V has a form like (14.42):

$$\tfrac{1}{4}\lambda(\phi_1'^2 + \phi_2'^2 + \phi_3'^2 + \chi^2 + 2f\chi + f^2 - h^2)^2. \tag{14.49}$$

This contains the extra terms

$$\tfrac{1}{2}\lambda(f^2 - h^2)(\phi_1'^2 + \phi_2'^2 + \phi_3'^2 + \chi^2 + 2f\chi). \tag{14.50}$$

Because (14.47) holds, U has the form

$$\tfrac{1}{4}\bar{\lambda}(\phi_1'^2 + \phi_2'^2 + \phi_3'^2 + \chi^2 + 2f\chi)^2 + O[(|\phi|^2 - |F|^2)^3], \tag{14.51}$$

where we have allowed for $\bar{\lambda}$ to differ from λ by terms of order $\hbar$. Thus the fate of (14.50) is to be cancelled by other contributions, like those in fig. 14.3.

The procedure in practice, then, is simply to omit all tadpole graphs and to omit (14.50). The quadratic terms in (14.50) also have to be left out, since there are no quadratic terms in ϕ_i' in (14.51). At the same time, the corresponding contributions from ϕ self-energy graphs must be removed. To be precise, if there is a contribution

$$\delta_{ij}\Sigma(p^2)$$

to the $\phi_i'\phi_j'$ self-energy function, then

$$\delta_{ij}\Sigma(0)$$

must be subtracted from this, and $\Sigma(0)$ must be subtracted from the χ self-energy function. In general the rule is that any contribution (at $p^2 = 0$) to the Goldstone field self-energies is to be subtracted from the whole multiplet of Higgs fields.

Our practice in earlier chapters, of imposing (14.48) or setting $h = f$ in (14.49), is therefore justified, provided it is understood that tadpoles and Goldstone self-energies (at $p^2 = 0$) are subtracted out at the same time.

15

Symmetry-breaking and mass-differences

15.1 Calculability

The combination of renormalizability and local gauge-invariance is a powerful one. In a renormalizable theory, every type of integral which does not have a counter-term corresponding to it must be finite. In a local gauge theory the counter-terms are limited to be the difference of two gauge-invariant Lagrangians (chapter 14). It follows that some quantities which at first sight one would expect to be divergent are, in suitable cases, actually convergent and calculable. In certain models, some mass-differences are finite, and some features of the spontaneous symmetry-breaking may even be calculable. Exciting as these possibilities are, they have not at present been incorporated into any realistic model. We shall describe two simple, but artificial, examples, in order to illustrate the principles involved.

15.2 Masses

We restrict ourselves to fermion masses. In a spontaneously broken symmetry, the masses in a multiplet are equal before the symmetry-breaking takes place (and fermion masses are zero if the symmetry is chiral). Mass-differences (and any mass at all in the chiral case) are caused by the coupling of the multiplet to the Higgs field, as the example in (8.28) illustrates. Yukawa interactions of the form

$$\bar{\psi}\phi\psi \tag{15.1}$$

are renormalizable, and so they must appear in the Lagrangian with arbitrary coupling constants. But interactions of the form

$$\bar{\psi}\phi^n\psi \quad (n > 1) \tag{15.2}$$

can appear only in the generating functional Γ (see §10.4) not in the Lagrangian itself; so they have finite, calculable coefficients.

Mass terms
$$\bar{\psi}F\psi \tag{15.3}$$

coming from (15.1) are arbitrary, but terms

$$\overline{\psi}F^n\psi \quad (n > 1) \tag{15.4}$$

from (15.2) are calculable. Useful results are obtained if (15.4) is not equivalent to (15.3).

One of the simplest cases occurs if (15.1) is forbidden by the discrete symmetry $\phi \to -\phi$. The phenomenon was first noted in a model of 't Hooft (1971 a). Let G be the (non-chiral) group $O(3)$, and let the Higgs field $\boldsymbol{\varphi}$ and the fermion $\boldsymbol{\psi}$ each be triplets. The fermions have, to zeroth order, a common mass m due to the invariant mass-term

$$-m\overline{\boldsymbol{\psi}}\cdot\boldsymbol{\psi}. \tag{15.5}$$

There is no Higgs–fermion coupling in the Lagrangian, because invariance under $\boldsymbol{\varphi} \to -\boldsymbol{\varphi}$ is imposed. But in Γ the vertex

$$-\gamma\overline{\boldsymbol{\psi}}\cdot\boldsymbol{\varphi}\,\boldsymbol{\varphi}\cdot\boldsymbol{\psi} \tag{15.6}$$

occurs, and, if $\mathbf{F} = (0, 0, f)$, gives a mass contribution

$$-\gamma f^2\overline{\psi}_3\psi_3. \tag{15.7}$$

The coefficient γ is calculable from the (convergent) graphs (a) and (b) of fig. 15.1 (with the external fermions on mass-shell).

In the Landau gauge ($\xi = 0$ in (6.19)), the graph (a) in fig. 15.1 is zero, because the $F\phi W$ vertices each contribute a factor k (the loop momentum) which annihilates the transverse W-propagator. Graph (b) (and higher-order graphs) give

$$f^2\frac{\partial}{\partial f^2}(\gamma f^2) = M^2\frac{\partial}{\partial M^2}\Sigma, \tag{15.8}$$

where Σ is the self-energy given by graph (c). Standard techniques give

$$i\Sigma = g^2(2\pi)^{-4}\int d^4k\overline{\psi}_3[4m + 2\gamma\cdot(k-p)]\psi_3\,[(k-p)^2 - m^2]^{-1}(k^2 - M^2)^{-1},$$

so that

$$\frac{\partial}{\partial M^2}\Sigma = -4img^2(2\pi)^{-4}\int_0^1 dx(1-x^2)\int d^4k[k^2 - x^2m^2 - (1-x)M^2]^{-3},$$

$$\gamma f^2 = -mg^2(8\pi^2)^{-1}\int_0^1 dx(1+x)\ln[1 + M^2(1-x)/(mx)^2]. \tag{15.9}$$

Thus, from (15.7), the neutral member of the fermion triplet is lighter than the charged ones.

No successful application of this idea has been made in a realistic model. To calculate the mass of the electron, it would have to be a

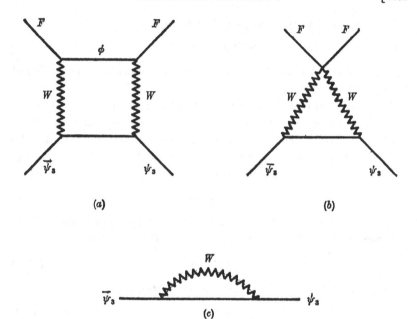

FIGURE 15.1. Graphs for calculating finite fermion
mass-difference.

member of a multiplet containing at least one other lepton (the muon?)
with a zeroth order mass (note that (15.9) is proportional to m).
Attempts in this direction have met with difficulties (Weinberg 1973b,
Mohapatra 1974, Georgi and Glashow 1973b, Eliezer 1974).

To calculate hadron electromagnetic mass-differences, one might
first calculate quark mass-differences. The effect of strong interactions
on the weak and electromagnetic interactions has to be considered
with care (see §18.4, and Weinberg 1973c). For other approaches see
Dicus and Mathur (1973), Freedman & Kummer (1973), Duncan and
Schattner (1973).

15.3 Calculation of symmetry-breaking

In the Salam–Weinberg theory (chapter 8), as in most simple models,
the spontaneous symmetry-breaking is controlled in effect by the
Lagrangian, in which the parameters may be chosen arbitrarily. It is
possible to construct examples where this is not the case.

Write the effective potential of (14.44) in the form

$$U(\phi) = V(\phi) + U_1(\phi), \tag{15.10}$$

where V is in the renormalized Lagrangian and U_1 represents higher-order contributions. In the Salam–Weinberg model, U and V are each functions of the single invariant $\phi^\dagger \phi$; and so the presence of U_1 makes no significant difference to the nature of the solution of (14.47).

Suppose, however, that $V(\phi)$ is invariant under a group $\bar{G}$ bigger than G. Then the stationary condition (14.48) on V is insufficient to determine F (up to an arbitrary transformation of G, which is always possible), and so the presence of U_1 is important. This situation arises if there are invariants constructed from ϕ which are of higher degree than the fourth (and which cannot be expressed in terms of lower degree invariants); for these invariants occur in U_1 but not in V.

We discuss a single example. Choose G to be $SU(3)$, and ϕ to transform as the octet representation, written as a traceless, Hermitian, 3×3 matrix. There are two independent invariants $\mathrm{Tr}(\phi^2)$ and $\mathrm{Tr}(\phi^3)$. The function V must have the form

$$V = a\,\mathrm{Tr}(\phi^2) + b[\mathrm{Tr}(\phi^2)]^2 + c\,\mathrm{Tr}(\phi^3), \tag{15.11}$$

and has stationary values $\phi = F$ only if

$$F^2 = \alpha F + \beta I. \tag{15.12}$$

Octets satisfying (15.12) have two equal eigenvalues, and they are called 'charges' (Michel and Radicati 1971). By an $SU(3)$ transformation, one can obtain

$$F \propto \lambda_8 = 3^{-\frac{1}{2}} \begin{pmatrix} 1 & 0 & 0 \\ 0 & 1 & 0 \\ 0 & 0 & -2 \end{pmatrix}. \tag{15.13}$$

The little-group G_F in this case is $SU(2) \times U(1)$. In general, higher-order terms in U_1 cannot produce a stationary point different from (15.13), for they cannot cancel the zeroth order term

$$3c[F^2 - \tfrac{1}{3}\mathrm{Tr}(F^2)]$$

arising from $\partial V / \partial \phi$.

Nothing interesting has been achieved so far. Suppose, however, we adjoin to G the discrete transformation $\phi \to -\phi$; so that $c = 0$ in (15.11). Then V is invariant under the group $\bar{G} = O(8)$, and the stationary condition on V fixes $\mathrm{Tr}(F^2)$ but not $\mathrm{Tr}(F^3)$. In this case, the higher order terms U_1 (which depend upon $[\mathrm{Tr}(\phi^3)]^2$ in general) might define a true minimum and determine $\mathrm{Tr}(F^3)$. Without loss of generality, one could choose

$$F \propto \cos\gamma\,\lambda_8 + \sin\gamma\,\lambda_3 \tag{15.14}$$

and γ would be *calculable* from U_1. The little-group G_F of (15.14) is $U(1) \times U(1)$.

The two alternatives we have described have different patterns of masses for the components of ϕ. In the first case, (15.13), ϕ_4, ϕ_5, ϕ_6, ϕ_7 (writing $\phi = \lambda_\alpha \phi_\alpha$) are Goldstone bosons and ϕ_1, ϕ_2, ϕ_3, ϕ_8 are Higgs particles (since $\lambda_1, \lambda_2, \lambda_3, \lambda_8$ commute with λ_8). In the second case (15.14), only ϕ_3, ϕ_8 are Higgs particles. But, whereas the field

$$\cos\gamma\,\phi_8 + \sin\gamma\,\phi_3 \qquad (15.15)$$

gets a mass from V in the usual way, the orthogonal field

$$-\sin\gamma\,\phi_8 + \cos\gamma\,\phi_3 \qquad (15.16)$$

acquires a mass only from U_1. Its (mass)^2 is therefore $O(\hbar)$ and is calculable. Such a particle has been termed by Weinberg (1973b) a 'pseudo–Goldstone–boson'.

A general treatment of pseudo–Goldstone–bosons is easily made by replacing G by $\bar{G}$ in the work following (6.34). Let $\bar{v}_A$ $(A = 1, \ldots, \bar{N})$ be the matrix representation, under which ϕ transforms, of the generators of $\bar{G}$. Components of ϕ in the space spanned by

$$\bar{v}_A F \quad (A = 1, \ldots, \bar{N}),$$

but not in the space spanned by (6.34), are pseudo–Goldstone–bosons.

15.4   One-loop contributions to the effective potential

In order to explore further the possibilities mentioned above, we must learn how to calculate the effective potential, at least to order $\hbar$. To simplify as much as possible, fermion fields are omitted.

The choice of gauge-fixing term presents a problem. The 't Hooft gauge terms (6.32) and (8.34) seem to be inconvenient for a number of related reasons:

(i) They contain the very quantity F which we are using U to calculate. Therefore the effective potential is a function $U(F, \phi)$, and the tadpole condition (14.47) becomes

$$\left[\frac{\partial U(F, \phi)}{\partial \phi}\right]_{\phi = F} = 0.$$

(ii) $U(F, \phi)$ is not, in general, invariant under the global form of the original representation of the group acting on ϕ alone, because the gauge-fixing term breaks this invariance. (U must be invariant under *some* realization of G acting on ϕ, because Γ and therefore U satisfy the Ward identities (12.27), (12.28). But the transformation law may involve F, and not be the original one.)

(iii) 't Hooft's gauge-fixing term adds a function of ϕ to the Lagrangian. It is not clear whether this should be included in U when U is minimized.

For all these reasons, it seems to be simplest to abandon 't Hooft's gauges, and use instead the gauge-fixing term

$$-\tfrac{1}{2}\xi^{-1}(\partial \cdot W^{\alpha})^2. \tag{15.17}$$

This leaves W-ϕ' mixing-terms uncancelled. It is not difficult to diagonalize combined (W, ϕ') quadratic terms in the Lagrangian; but in fact, for the calculation of U it is possible to work with the original field ϕ instead of ϕ'.

What types of simple closed loop can be drawn? In the gauges (15.17), there are no spurion closed-loop contributions to $U(\phi)$, since ϕ does not occur in the spurion Lagrangian (11.40). Assuming invariance under $\phi \to -\phi$, only vertices with an even number of ϕ-lines exist. We then distinguish three types of loop, exemplified in fig. 15.2, where solid lines represent ϕ-propagators and zig-zag lines W-propagators. Loops of type (a) contain ϕ-lines only, those of type (b) W-lines only. All mixed loops are included in type (c).

Graphs of type (a) are generated by the $V(\phi)$ part of the Lagrangian and so are invariant under the bigger group $\bar{G}$. They are irrelevant to the calculation of $\bar{G}$-breaking, and they can be ignored. If we specialize to the Landau limit $\xi \to 0$ in (15.17), graphs of type (c) are zero. This is because the $\phi\phi W$ vertices give factors of k (the loop momentum), which, contracted with the transverse propagator ($\xi = 0$ in (6.19)), gives zero. The result for the remaining graphs of type (b) is actually the same as if we had taken the Landau limit, $\xi \to 0$, in the 't Hooft term (6.32).

We now compute the sum of all graphs of type (b) in fig. 15.2. We treat the general case of §6.5 since it is as easy as the particular $SU(3)$ model of §15.3. The vertices are given by the term

$$\tfrac{1}{2}g^{\alpha}g^{\beta}(\tilde{\phi}\tilde{v}^{\alpha}v^{\beta}\phi)g^{\lambda\nu}W^{\alpha}_{\lambda}W^{\beta}_{\nu} = \tfrac{1}{2}\mathcal{M}^{\lambda\nu}_{\alpha\beta}W^{\alpha}_{\lambda}W^{\beta}_{\nu}, \tag{15.18}$$

and the propagator by

$$(-g_{\lambda\nu} + k_{\lambda}k_{\nu}/k^2)(k^2 + i\epsilon)^{-1}\delta^{\alpha\beta} \equiv P^{\alpha\beta}_{\lambda\nu}(k^2 + i\epsilon)^{-1}. \tag{15.19}$$

No mass term appears in this propagator, since we are using ϕ not ϕ'. If one sets $\phi = F + \phi'$ in (15.18), the terms $\tilde{F}\tilde{v}_{\alpha}v_{\beta}F$ in (15.18) generate the masses in the denominators when the sum is taken over all loops.

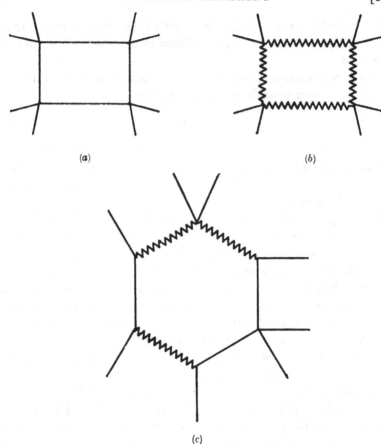

FIGURE 15.2. Graphs contributing to the effective potential
(straight lines, Higgs field; zig-zag lines, vector field).

In terms of these definitions, the sum of all loops like (b) is

$$-\mathrm{i}U(\phi) = \int \mathrm{d}^4k \sum_{n=1}^{\infty} n^{-1} \mathrm{Tr}\left[(\mathscr{M}P/k^2)^n\right]$$

$$= \int \mathrm{d}^4k \sum_{n=1}^{\infty} n^{-1} \mathrm{Tr}\left[(\mathscr{M}/k^2)^n P\right]$$

$$= -\int \mathrm{d}^4k \,\mathrm{Tr}\left\{P[\ln\left(I - \mathscr{M}/k^2\right)]\right\} + (\text{constant})$$

$$= -\int \mathrm{d}^4k \,\mathrm{Tr}\left\{P[\ln\left(\mathscr{M} - k^2 I\right)]\right\} + (\text{constant})$$

$$= -\tfrac{3}{4}\mathrm{d}^4k \int \mathrm{Tr}\left[\ln\left(\mathscr{M} - k^2 I\right)\right] + (\text{constant}). \qquad (15.20)$$

In these expressions, the traces include Lorentz and $SU(3)$ indices. The combinatorial factor n^{-1} comes about because cyclic permutations of the vertices around a closed loop do not change them.

In each term in the series in (15.20), one can make the usual Wick rotation through 90° anticlockwise in the complex k_0-plane. Then the final form of (15.20) becomes

$$U(\phi) = \tfrac{3}{4}\pi^2 \int_0^\infty k^2 \, d(k^2) \, \text{Tr} \, [\ln(k^2 I + \mathcal{M})] + (\text{constant}). \quad (15.21)$$

This integral, like the first two terms in the series in (15.20), is divergent. Integrating up to an upper limit Λ^2, (15.21) gives

$$\tfrac{3}{8}\pi^2 \, \text{Tr} \, [\mathcal{M}^2 \ln \mathcal{M}] + (\text{const.}) \, \text{Tr} \, (\mathcal{M}^2) + (\text{const.}) \, \text{Tr} \, (\mathcal{M}) + (\text{const.}). \quad (15.22)$$

The constants in (15.22) are divergent as $\Lambda^2 \to \infty$, but they multiply terms of degree four or less in ϕ and so they have the same structure as $V(\phi)$. They do not break the $\bar{G}$ symmetry. Thus

$$U_1(\phi) = \tfrac{3}{8}\pi^2 \, \text{Tr} \, [\mathcal{M}^2 \ln \mathcal{M}] + (\text{terms invariant under } \bar{G}), \quad (15.23)$$

where $\mathcal{M}$ is defined in (15.18). The trace can be computed by diagonalizing $\mathcal{M}$.

The form (15.23) was derived by Weinberg (1973b). A general treatment of the effective potential has been worked out by Jackiw (1974).

Finally, let us return briefly to the particular $SU(3)$ model of §15.3. It is found (Coleman and Weinberg 1973) that, as a function of γ defined in (15.14), the only minimum of U is at $\gamma = 0$. The existence of a stationary value at this particular value of F (satisfying (15.12)) is guaranteed by a general theorem (Michel and Radicati 1971). There seems to be no reason in general why there should not be other, more interesting, minima. It is disappointing that, in this model at least, one does not occur.

15.5 Beyond perturbation theory

Up to this point, we have made the usual assumption of perturbation theory that all terms $O(\hbar)$ are smaller than any term $O(\hbar^0)$. Then U_1 cannot appreciably change the minima of V; and U_1 is only important in cases where V does not have a true minimum (like (15.11) when $c = 0$).

This assumption is not obviously justified when there is more than one coupling constant. Some authors have tried other assumptions. We briefly mentioned one example, chiefly to emphasize the assumption made.

Coleman and E. Weinberg (1973) assume that $V(\phi)$ is of order $\hbar g^4$. They also impose the condition that the renormalized mass-term in V is zero, so that V has quartic terms only. Then V has no minimum except the symmetry-preserving one at $\phi = 0$, but U may have another minimum. The condition for U to have a symmetry-breaking minimum determines the strength λ of the quartic coupling in V in terms of g. The position of the minimum remains an arbitrary parameter because of the arbitrariness in the definition of the renormalized value of λ.

16

Higher-order corrections

16.1 General considerations

The main point of the gauge theories of weak interactions is that they are renormalizable, so that useful higher-order calculations can be made. The questions immediately arise: what are the results of such calculations, how large are they, and can they be tested experimentally? In particular, is the effective expansion parameter g^2 (which is order e^2 in the Salam–Weinberg model of chapter 8), or is it

$$g^2 E^2/M_W^2 \sim G_W^2 E^2, \qquad (16.1)$$

where E is some relevant energy (such as the nucleon, muon or electron mass)? Are there parity-violating effects of order g^2?

It is possible to suggest answers to these questions by the following arguments. We restrict ourselves to moderate energy processes in which no real W or Z particles (or Higgs particles) are emitted or absorbed. Then one-loop Feynman graphs can be divided into three sorts according as the number of W or Z internal lines is 0, 1, or 2. (Higgs particles play no very important role in deciding orders of magnitude.) The first sort consists of ordinary electromagnetic corrections.

Consider next the case of just one W or Z internal line. The Feynman integral has the form

$$g^2 \int \mathrm{d}^4 k f_1(k) \, (k^2 - M^2)^{-1}. \qquad (16.2)$$

If $f_1 \sim (k^2)^{-2}$, (16.2) is convergent and is of order

$$(g^2/M^2) \ln (M^2) \sim G_W \ln (M^2), \qquad (16.3)$$

although the logarithm may be absent in particular cases. This is a truly weak term, although its actual magnitude depends upon the relevant energy scale both in the overall factor and in the logarithm.

If $f_1 \sim (k^2)^{-1}$ in (16.2), the integral is divergent. In a renormalizable theory, the infinite part is to be subtracted, but a non-trivial finite part may remain. Suppose, for example, there are a number of vector mesons, so that M^2 in (16.2) is a matrix $(M^2)_{\alpha\beta}$. Then from (13.14), (16.2) contributes a finite part proportional to

$$g^2 (\ln M^2)_{\alpha\beta}. \qquad (16.4)$$

This has a more general structure than the infinite part (which is proportional to $\delta_{\alpha\beta}$); and so might produce important symmetry-breaking effects. (Indeed, §15.2 is really a case in point.) We do not encounter instances of this type in the remainder of the chapter, but we return to (16.4) in §18.4.

Finally, consider graphs containing two W or Z lines. They contribute integrals of the form (where $f_2 \sim (k^2)^{-1}$)

$$g^4 \int d^4 k f_2(k) (k^2 - M^2)^{-2} \sim g^4 (M^2)^{-1} \sim g^2 G_W. \qquad (16.5)$$

Weak corrections to weak processes, therefore, are expected generally to be of relative order g^2.

In the following sections, we give examples which show how the above rough general arguments have to be refined in particular cases.

16.2 The muon magnetic moment

The purely electromagnetic prediction for the muon anomalous magnetic moment $\frac{1}{2}(g-2)$ is (Combley and Picasso 1974)

$$(11\,658.1 \pm 0.2) \times 10^{-7} \qquad (16.6)$$

and experiment gives $(11\,661.6 \pm 3.1) \times 10^{-7}. \qquad (16.7)$

Hadronic contributions (which depend, however, on assumptions about the ratio (13.43)) are estimated to be less than 10^{-7}. From the arguments of §16.1, we expect a weak contribution of order

$$G_W m_\mu^2 \simeq 10^{-7}. \qquad (16.8)$$

A logarithmic factor $\ln (M^2/m_\mu^2) \simeq 12 \qquad (16.9)$

might possibly make (16.8) significant, so a more careful calculation is necessary.

The graphs are shown in fig. 16.1, where (a) is purely electromagnetic and leads to (16.6). We use the gauge $\xi = 1$ in (6.19). Each graph is initially logarithmically divergent; but when the part proportional to the magnetic-moment term.

$$m_\mu \bar{u} \sigma_{\lambda\nu} u q^\nu$$

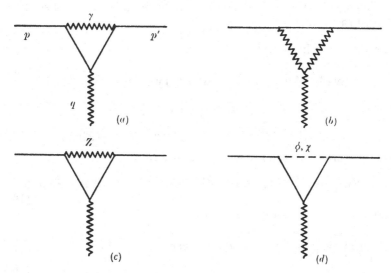

FIGURE 16.1. Muon magnetic moment graphs (solid line, leptons; broken line, Higgs fields; zig-zag line, vector fields).

(u is the muon spinor) is extracted, it is convergent. So we do not have to worry about renormalization.

Graph (b) has two W lines and therefore according to (16.5) has no logarithmic factor (16.9). Graph (d) includes contributions from the Higgs particle χ and from the Goldstone fields ϕ_1, ϕ_2, ϕ_3. In the latter case, the couplings to the muons are, from (8.28), proportional to

$$G_W^{\frac{1}{2}} m_\mu \sim g m_\mu / M_W$$

(see (8.30) and (8.31)); so the contribution is reduced by a factor m_μ^2 / M_W^2 relative to graph (c). For the Higgs particle graph, the mass in the propagators is M_χ not M_W; so the overall contribution has a factor m_μ^2 / M_χ^2 relative to graph (c). Little is known about the value of M_χ (see §9.5), but, even if M_χ were less than m_μ, the contribution from the Higgs particle graph would only be of the same order as that from graph (c) (see Jackiw and Weinberg 1972).

It is not obvious whether the important graph, (c), has a logarithmic factor (16.9). We discuss this graph in a little more detail. In the calculation of the magnetic moment, there can be no interference between the vector and the axial pieces of the Z current (8.14) (such interference would give an electric dipole moment which is forbidden by time-reversal invariance).

We consider the axial current contribution, for which the relevant part of (8.14) is

$$-\tfrac{1}{4}\bar{\mu}\gamma_\lambda\gamma_5\mu. \qquad (16.10)$$

The Feynman integral is

$$ig^2e(16\cos^2\theta_W)^{-1}(2\pi)^{-4}\int d^4k\,\bar{u}(p)\,\gamma_\nu\gamma_5[\gamma\cdot(p-k)-m_\mu]\gamma_\lambda$$

$$\times\,[\gamma\cdot(p'-k)-m_\mu]\gamma^\nu\gamma_5 u(p')$$

$$\times\,\{[(p-k)^2-m_\mu^2]\,[(p'-k)^2-m_\mu^2]\,[k^2-M_Z^2]\}^{-1}. \qquad (16.11)$$

The numerator in the integral simplifies to

$$-2\bar{u}(p)\,[\gamma\cdot(p'-k)\,\gamma_\lambda\gamma\cdot(p-k)+m_\mu^2\gamma_\lambda-2m_\mu(p_\lambda+p'_\lambda-2k_\lambda)]\,u(p') \qquad (16.12)$$

and the denominator can be written

$$2\int_0^1 dx\int_0^x dy\,\{[k-yp-(x-y)p']^2+y(x-y)\,q^2-x^2m_\mu^2-(1-x)\,M_Z^2\}^{-3}. \qquad (16.13)$$

After defining $k = k' + yp + (x-y)\,p'$,

the k'-integration is done, to give

$$-eg^2(2^7\pi^2\cos^2\theta_W)^{-1}\int_0^1 dx\int_0^x dy\,N(x,y)\,[x^2m_\mu^2+(1-x)\,M_Z^2-y(x-y)\,q^2]^{-1}$$

$$+\,(\text{terms proportional to }\bar{u}\gamma_\lambda u),\quad (16.14)$$

where

$$N = \bar{u}\{[(1-x)\,m_\mu-(1-x+y)\,\gamma\cdot q]\,\gamma_\lambda[(1-x)\,m_\mu+(1-y)\,\gamma\cdot q]$$

$$-\,2m_\mu(1-x)\,(p+p')_\lambda\}\,u. \qquad (16.15)$$

Using the identity

$$\bar{u}(p+p')_\lambda u = 2m_\mu\bar{u}\gamma_\lambda u+\bar{u}\sigma_{\nu\lambda}u q^\nu \qquad (16.16)$$

and again neglecting terms proportional to $\bar{u}\gamma_\lambda u$, (16.15) may be replaced by (($x-2y$) gives zero on integration)

$$-2m_\mu(1-x)\,(4-x)\,\bar{u}\sigma_{\nu\lambda}u q^\nu. \qquad (16.17)$$

For the anomalous magnetic moment, we set $q^2 = 0$ in (16.14). Because of the factor $(1-x)$ in (16.17), it is also possible to approximate (16.14) by setting $m_\mu^2 = 0$ without thereby creating a divergence at $x = 1$. This gives a contribution to $(g-2)\,e/2m_\mu$ of amount

$$\tfrac{5}{3}g^2(2^5\pi^2\cos^2\theta_W)^{-1}\,(m_\mu/M_Z)^2\,(e/2m_\mu). \qquad (16.18)$$

The contribution from the vector coupling of the Z-particle is

$$\tfrac{1}{3}g^2(1-4\sin^2\theta_W)^2\,(2^5\pi^2\cos^2\theta_W)^{-1}\,(m_\mu/M_Z)^2\,(e/2m_\mu). \qquad (16.19)$$

Combining (16.18), (16.19) and the W-meson contribution from fig. 16.1(b), the total result for $(g-2)$ comes out to be (Jackiw and Weinberg 1972)

$$(G_W m_\mu^2/\sqrt{2\pi^2})(\tfrac{4}{3}\sin^4\theta_W - \tfrac{2}{3}\sin^2\theta_W + \tfrac{1}{2}), \qquad (16.20)$$

which is of order 10^{-8} and so is negligible in comparison with (16.6) and (16.7). The important point is that there turns out, because of the $(1-x)$ factor in (16.17), to be no logarithm like (16.3).

The above calculations are equally applicable to the electron. The experimental result is more accurate than (16.7) by a factor of 100. But the weak corrections are smaller than (16.20) by a factor

$$(m_e/m_\mu)^2 \simeq 10^{-4},$$

and so they are not significant.

16.3   Corrections to muon decay

Four of the many graphs are shown in fig. 16.2. In old-fashioned vector meson theory (or in the unitary gauge, $\xi \to \infty$, of the Salam–Weinberg model), the photon graphs like (a) and (b) gave logarithmically divergent contributions (although in the Fermi direct coupling theory the photon corrections happened to be finite). The corrections to the electron's spectrum, however, were finite and quite significant. Two things tend to amplify the corrections. One is a factor $\ln(m_\mu/m_e)$. The other is the presence of infra-red divergences from virtual photons. These require the inclusion of the soft photon decay

$$\mu^- \to e^- \nu_\mu \bar\nu_e \gamma$$

to cancel them. The predictions then depend logarithmically upon the resolution with which the electron's energy is measured, since this limits the energy of a possible soft-photon. (See Bailin 1971 for a review of the calculations before the invention of gauge theories.)

The inclusion of the graphs of the Salam–Weinberg models, like (c) in fig. 16.2, alter the corrections to the spectrum only by insignificant terms of order

$$(g^4/M_W^4)\, m_\mu^2 \sim G_W^2 m_\mu^2,$$

as expected on dimensional grounds. Thus gauge theories have nothing new to say about the spectrum.

In the Salam–Weinberg theory, the correction to the rate is finite, provided it is expressed in terms of properly renormalized quantities.

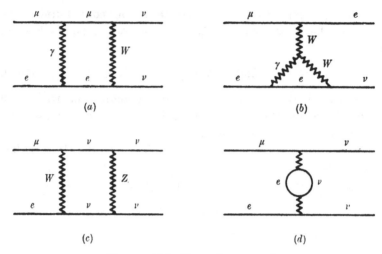

FIGURE 16.2. Muon decay graphs.

But the question is what to compare the rate with? The renormalized parameters of the model are fixed by e, M_W, M_Z, M_χ, m_μ and m_e. *In principle*, these are all measurable, and the muon decay rate then predictable (Ross 1973, Appelquist, Primack and Quinn 1973). To this prediction, graphs like (d) in fig. 16.2 contrbute terms of order

$$G_W g^2 \ln (M_W^2/m_\mu^2). \qquad (16.21)$$

This is an example where logarithmic enhancement does take place. Unfortunately, experimental confirmation is a distant prospect.

It is of more immediate interest to compare the rate of muon decay with, say, neutron decay. This comparison is one way of determining the Cabibbo angle θ_C, since the neutron decay amplitude is proportional to $\cos \theta_C$. Such a comparison requires a theory of hadronic weak interactions and techniques for handling the strong interactions.

A calculation has been made (Angerson 1974) treating the proton and neutron as if they were leptons (that is, ignoring the strong interactions). A logarithmic factor

$$G_W(3\alpha/4\pi) \ln (M_Z^2/m_P m_e) \qquad (16.22)$$

was found, giving a correction of about 5 % increasing the neutron decay rate with respect to the muon decay rate. This result is just as if M_Z acted as a cut-off to the ordinary electromagnetic corrections in Fermi direct-coupling weak interaction theory. The correction is of the right sign and order of magnitude to reconcile the two decay rates with a Cabibbo angle.

However, it is difficult to justify this type of approximation. The logarithm in (16.22) comes from virtual momenta of the order of M_Z, which are presumably associated with structure deep inside the nucleon. It is perhaps a better starting point to calculate corrections to quark beta decay, neglecting strong interactions. Certainly the corrections are sensitive to the charges of the quarks (Abers, Dicus, Norton and Quinn 1968, Preparata and Weisberger 1968, Lee 1972, Mohapatra and Sakakibara 1974, Sirlin 1974).

16.4   Corrections to forbidden processes

As explained in § 9.1, any model of hadronic weak interactions must be constructed to forbid neutral strange current processes, at least in lowest order. We take as one example the observed branching ratio

$$\frac{K_L^0 \to \mu^+\mu^-}{K_L^0 \to \text{all}} \simeq 10^{-8}. \tag{16.23}$$

The mass-difference $\qquad \Delta m_K = m_{K_L} - m_{K_S} \qquad$ (16.24)

is observed to have a magnitude

$$\Delta m_K / m_K \simeq 0.5 \Gamma_{K_S} / m_K \simeq 7 \times 10^{-15}. \tag{16.25}$$

This tells us that Δm_K is of order G_W^2, and certainly not of order G_W.

It is important to verify that higher-order corrections are consistent with these very small ratios. Fig. 16.3 shows a typical graph (a) generating a strange neutral current, and therefore contributing indirectly to (16.23). Graph (b) gives a $\Delta S = 2$ amplitude contributing indirectly to the transition $\qquad K^0 \leftrightarrow \bar{K}^0$

and thus to (16.24).

Each graph individually is of order (16.5), and therefore about 10^{-2} of a lowest-order contribution (if one existed). Such an order of magnitude is certainly too large to be consistent with (16.23) or (16.25). However, the cancellation between p and c quarks, which was used in § 9.1 to eliminate neutral strange currents, operates here also. After the cancellation, the order of magnitude of the contribution from the graphs (a) is

$$\sin \theta_C g^4 M_W^{-4} (m_c^2 - m_p^2) \ln (M_W^2), \tag{16.26}$$

which gives the ratio (16.23) of order

$$[\sin \theta_C G_W (m_c^2 - m_p^2) \ln (M_W^2)]^2. \tag{16.27}$$

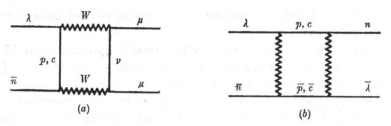

FIGURE 16.3. Quark graphs changing strangeness.

Similarly, the graph (b) gives (16.25) to be proportional to

$$\sin^2\theta_C\, G_W^2(m_c^2 - m_p^2)^2\,(m_c^2 + m_p^2)^{-1}\ln{(M_W^2)}. \qquad (16.28)$$

The estimate (16.27) is comfortably consistent with (16.23). Comparison of (16.28) with (16.25) is more delicate, requiring the insertion of a factor of dimension (mass)^2. By making assumptions about the matrix-elements of the quark operators between kaon states, Lee, Primack and Treiman (1973) conclude that $(m_c - m_p)$ is less than 1 GeV.

These questions have recently been carefully re-examined by Gaillard and Lee (1974b). They also discuss decays expected to be of order $e^2 G_W$, like $K^{\pm} \rightarrow \pi^{\pm} e^+ e^-$.

17

CP and *T* violation

17.1 Summary of the evidence

From the existence of the decay modes $K_L^0 \to 2\pi$ and from the observed charge asymmetry in $K^0 \to \pi^\pm l^\mp \bar{\nu}(\nu)$, it is known that *CP* conservation is not exact. There is also indirect evidence for time-reversal violation, but no evidence for *CPT* violation.

The states of definite life-time K_S^0 and K_L^0 are expressible in the form

$$[2(1+|\epsilon|^2)]^{-\frac{1}{2}}\,[(1+\epsilon)\,K^0 \pm (1-\epsilon)\,\bar{K}^0], \qquad (17.1)$$

where K^0, $\bar{K}^0$ are *CPT* eigenstates. The phase of ϵ is given approximately by

$$\arg \epsilon \simeq \tan^{-1}[2(m_{K_L}-m_{K_S})/\Gamma_{K_S}] \simeq 45^0, \qquad (17.2)$$

and the modulus is determined from the charge asymmetry results:

$$|\epsilon| \simeq 2 \times 10^{-3}. \qquad (17.3)$$

Equations (17.1), (17.2) and (17.3) are capable of fitting all the data, without invoking any *CP*-violation in the decay amplitudes themselves. This fact is most neatly explained by the superweak theory of Wolfenstein (1964), according to which the *CP*-violating amplitude is of order $10^{-9}\,G_W$ and changes strangeness by 2 (or more). It can, therefore, contribute in first order to the K^0-$\bar{K}^0$ mixing matrix, where its influence is amplified by the small K_L^0-K_S^0 mass-difference. It is too small to have detectable effects elsewhere in physics.

An alternative model is the milliweak one, in which the *CP*-violating amplitude is of order $10^{-3}\,G_W$ and changes strangeness by 1. The observed effects are then of second order. On this model, it is not obvious why *CP*-violating effects in the decay amplitudes do not complicate the predictions from (17.1). Accurate measurements of the neutron electric dipole moment may distinguish between superweak and milliweak models.

17.2 Spontaneous breaking of *CP* and *T*

Models have been constructed in which *CP* and *T* violation is attributed to spontaneous breaking of the symmetries. These models produce a milliweak situation, and they provide some sort of explanation of the smallness of the violation. We briefly describe the general idea of the models.

Given a Dirac field ψ, the Hermitian scalar and pseudoscalar

$$\bar{\psi}\psi, \quad i\bar{\psi}\gamma_5\psi \tag{17.4}$$

have opposite *CP* and *T* transformation properties. (In this respect, they are unlike the vector and axial-vector $\bar{\psi}\gamma_\lambda\psi, \bar{\psi}\gamma_\lambda\gamma_5\psi$.) This is the key to *CP* violation by a Higgs field.

The simplest model uses a Higgs field ϕ and a Lagrangian containing

$$m\bar{\psi}\psi + ia\bar{\psi}\gamma_5\psi\phi, \tag{17.5}$$

where a is a (real) coupling constant. This conserves *CP* if ϕ is assigned $CP = -1$. But if spontaneous symmetry-breaking gives ϕ a non-zero vacuum-expectation-value f, (17.5) may be rewritten

$$(m^2 + f^2 a^2)^{\frac{1}{2}}\bar{\psi}'\psi' + a\bar{\psi}'(\sin\alpha + i\gamma_5\cos\alpha)\psi'\phi', \tag{17.6}$$

where $\qquad \phi = f + \phi', \quad \psi' = \exp(\tfrac{1}{2}i\gamma_5\alpha)\psi, \quad \tan\alpha = af/m. \tag{17.7}$

(Vector or axial-vector interactions are unaffected by the transformation from ψ to ψ'.) The *CP* violation is now caused by the exchange of ϕ' particles. Since the coupling of Higgs fields is usually rather small (see, for example, (8.28)), it is possible to arrange for the *CP* violation to be of roughly milliweak magnitude.

The situation is different for a complex Higgs field. Its *CP* transformation law contains an arbitrary phase which can be chosen so that the vacuum-expectation-value has $CP = +1$. With two complex Higgs fields, however, the vacuum-expectation-values may be out of phase, and so induce *CP* and *T* violation.

The considerations of §6.9 about high temperatures and number densities also apply to spontaneous *CP* violation. Kobsarev, Okun and Zeldovich (1974) argue that some models imply a domain structure for the universe which is incompatible with the observed isotropy of the microwave background radiation.

A lucid review of spontaneous *CP* violation is provided by T. D. Lee (1974). Other contributions are due to Mohapatra (1972), Pais (1973), Mohapatra and Pati (1975), Bailin and Love (1974).

It could be that a *CP* violating minimum of the effective potential U (see §14.6) appears only when the quantum corrections U_1 are included. This possibility is discussed by Zee (1974), Mohapatra (1974) and Georgi and Pais (1974).

18

Gauge theories and strong interactions

18.1 Requirements of a theory of strong interactions

If weak and electromagnetic interactions of hadrons are to be calculated in closed-loop calculations, some assumptions must be made about the strong interactions. There is little point in having a renormalizable theory of weak interactions unless it is assumed that strong interactions are renormalizable also.

When renormalizing the complete (strong, electromagnetic and weak) Lagrangian, *all* allowed renormalizable terms must be included. 'Allowed' here means constructable from the fields in the theory, and not forbidden by any of the symmetries of the complete Lagrangian. The question then arises whether there are parity violating terms in the strong-interaction parts of the complete Lagrangian. Of course, there may be such terms which *happen* to have very small coefficients; but a model carries more conviction if there is a natural explanation for the absence of parity violation in the strong interactions.

As a simple example, suppose among the hadron fields there are two spinors ψ_1, ψ_2 and a spin-0 field ϕ. In general, the couplings

$$\beta\bar{\psi}_1\psi_2\phi + \beta'\bar{\psi}_1\gamma_5\psi_2\phi \qquad (18.1)$$

are allowed and must be included. It may be possible to remove the apparent violation of parity in (18.1) by transformations of the form

$$\psi'_i = \exp(\theta_i\gamma_5)\psi_i \qquad (18.2)$$

(whether it is possible depends upon the remainder of the Lagrangian). If it is not so possible, a model allowing (18.1) does not naturally account for parity conservation in the strong interactions.

Which of the strongly-interacting fields have the weak and electromagnetic interactions? That is, which of them transform other than as singlets under the weak group G? The answer suggested by the quark-parton model (see §9.3(iv)) is: the quark fields. Certainly, if scalar or vector fields had any direct weak or electromagnetic interaction, it would be hard to understand the apparent success of the

[147]

quark-parton model. Spin-0 fields would not explain the ratio (9.25), and spin-1 fields would spoil scaling altogether.

These considerations suggest a general type of model to which we shall return in § 18.4. But first we mention two other models, in each of which the junction between weak and strong interactions is located, partly at least, in the Higgs fields. These models seem not to satisfy the criteria we have just mentioned, and they are certainly not complete theories. But each has an interesting merit, the first explaining chiral-symmetry-breaking, the second accommodating naturally the Cabibbo angle.

18.2 Higgs fields and breaking of chiral symmetry

In § 5.5 we touched upon the subject of $PCAC$ and the breaking of the chiral group $SU(2)_L \times SU(2)_R$ in the σ-model. In this model, a small explicit symmetry-breaking term (5.39)

$$-\epsilon\sigma(x) \tag{18.3}$$

was invoked in order to give pions their non-zero masses. The field σ is one component of a $(\tfrac{1}{2},\tfrac{1}{2})$ representation $(\sigma,\boldsymbol{\pi})$. Can this explicit symmetry-breaking be avoided?

The transformation properties of the $(\sigma,\boldsymbol{\pi})$ multiplet are displayed in the matrix

$$2^{-\frac{1}{2}}(\sigma+i\boldsymbol{\tau}\cdot\boldsymbol{\pi}) = \begin{pmatrix} 2^{-\frac{1}{2}}(\sigma+i\pi_3) & i\pi^+ \\ i\pi^- & 2^{-\frac{1}{2}}(\sigma-i\pi_3) \end{pmatrix} \tag{18.4}$$

on which $SU(2)_L$ and $SU(2)_R$ transformations operate from the right and left respectively. Thus, as far as $SU(2)_L$ is concerned, the doublet

$$\Pi = \begin{pmatrix} i\pi^+ \\ 2^{-\frac{1}{2}}(\sigma-i\pi_3) \end{pmatrix}$$

has identical transformation properties to the Higgs field doublet ϕ in (8.21) (we ignore strangeness in this section). Thus the interaction

$$-\epsilon'(\Pi^\dagger\phi + \phi^\dagger\Pi) \tag{18.5}$$

is invariant under $SU(2)_L$.

For small enough ϵ', the vacuum-expectation-values of ϕ and Π are little changed by (18.5), and are given approximately by (8.20) and (5.40). Since the former is so much bigger, the dominant effect of (18.5) is to produce a term
$$-\epsilon'f\sigma.$$

This simulates the required term (18.3) if

$$\epsilon' = \epsilon/f = m_\pi^2 f_\pi/f \simeq 3.6 \times 10^{-4} m_\pi^2. \qquad (18.6)$$

Thus the interaction (18.5) can reasonably be called weak, yet it triggers the substantial chiral-symmetry-breaking in strong interactions. This is because of the large value of f, which just reflects the weakness of weak interactions.

This type of model has been elaborated by Weinberg (1971 a), Palmer (1972), Weinstein (1973 a), Hagiwara and Lee (1973), Jones (1974). Bailin and Love (1974) discuss spontaneous CP violation (see § 17.2) in the framework of such a model.

18.3 Gauge theories of strong-interaction symmetries

Strong interactions are invariant under the isotopic-spin group $SU(2)$, and they show signs of a larger approximate symmetry $SU(3)$. Can these groups be connected with a Higgs-type gauge theory? The 1^- mesons ρ, ω, ϕ, K^* are naturally associated with a Yang–Mills field for the group $U(3)$ (since there are nine mesons). The difficulty is that, although the isotopic subgroup $SU(2)$ is unbroken in the strong interactions, the ρ-mesons are not markedly lighter than the other vector mesons.

Models which overcome this difficulty, each in essentially the same way, have been constructed by de Wit (1973) and Bars, Halpern and Yoshimura (1973). We outline the simple model of de Wit, which ignores chiral symmetries.

In this model, the strong-interaction Lagrangian is invariant under

$$U(3)_{\text{LOCAL}} \times U(3)_{\text{GLOBAL}}, \qquad (18.7)$$

in which neither factor is the familiar unitary symmetry group (which we will call $U(3)_{\text{PHYSICAL}}$). There is a nonet of Yang–Mills fields associated with $U(3)_{\text{LOCAL}}$. Higgs fields are introduced in a 3×3 matrix $K(x)$, transforming as

$$K(x) \to U(x) K(x) V, \qquad (18.8)$$

where $\dot{U}(x)$ is a matrix of the fundamental representation of $U(3)_{\text{LOCAL}}$ and V is a matrix of the fundamental representation of $U(3)_{\text{GLOBAL}}$.

By transformations of the form (18.8), the vacuum-expectation-value of K may be chosen to be proportional to the unit matrix. This breaks invariance under (18.7) but leaves invariance under the global $SU(3)$ subgroup defined by

$$U(x) = V^\dagger, \quad \det V = 1. \qquad (18.9)$$

This is identified with $SU(3)_{\mathrm{PHYSICAL}}$, which at this stage is an exact symmetry. Higgs' mechanism makes the vector mesons massive (a degenerate octet and a singlet). There is a nonet of Higgs particles, corresponding to the Hermitian part of $K(x)$.

The next step is to introduce weak and electromagnetic interactions, at the same time breaking the $SU(3)_{\mathrm{PHYSICAL}}$ symmetry. To do this, first of all extend (18.7) so that an $SU(2) \times U(1)$ subgroup of the $U(3)_{\mathrm{GLOBAL}}$ in (18.7) is made into a local group. This subgroup is identified with the group of the Salam–Weinberg model, and it is broken by the vacuum-expectation-value of a Higgs field ϕ, in the usual way.

The fields $K(x)$ transform under the $U(2)_{\mathrm{LOCAL}}$. The crucial question is: what simultaneous vacuum-expectation-values are possible for the two Higgs fields K and ϕ? De Wit answers this question by first introducing a 3×3 matrix C which simply defines the way the $U(2)_{\mathrm{LOCAL}}$ is embedded in $U(3)_{\mathrm{GLOBAL}}$. Thus

$$\tilde{\phi}(x) = C^\dagger \begin{pmatrix} 0 & 0 & \phi_1 \\ 0 & 0 & \phi_2 \\ 0 & 0 & 0 \end{pmatrix} C \qquad (18.10)$$

is defined to transform like $K(x)$ under (18.7). (We write here ϕ_1, ϕ_2 for the two complex components of the doublet ϕ.) The choice of C is arbitrary, and it could for instance be taken to be the unit matrix. But de Wit prefers to fix C indirectly by the condition that it should allow the vacuum-expectation-values of K and ϕ to have simple forms. He shows, in fact, that one can arrange

$$\langle 0| K(x) |0 \rangle = \begin{pmatrix} f_1 & 0 & 0 \\ 0 & f_2 & 0 \\ 0 & 0 & f_3 \end{pmatrix}, \qquad (18.11)$$

$$\langle 0| \phi(x) |0 \rangle = 2^{-\frac{1}{2}} \begin{pmatrix} 0 \\ f \end{pmatrix}, \qquad (18.12)$$

provided that C has the form

$$C = \begin{pmatrix} 1 & 0 & 0 \\ 0 & \cos\theta_{\mathrm{C}} & \sin\theta_{\mathrm{C}} \\ 0 & -\sin\theta_{\mathrm{C}} & \cos\theta_{\mathrm{C}} \end{pmatrix}. \qquad (18.13)$$

It is clear that θ_{C} is to be identified with the Cabibbo angle, and it is the great merit of de Wit's model that such an angle occurs naturally. The breaking of $SU(3)_{\mathrm{PHYSICAL}}$ is due to f_3 in (18.11) being different from f_1 and f_2. In principle this is an electromagnetic effect, since it goes away when weak and electromagnetic interactions are turned off. But it need not be small, because of the large magnitude of f in (18.12). In this respect, de Wit's model resembles that of §18.2.

There is nothing to make $f_1 = f_2$; and $f_1 \neq f_2$ causes a direct violation of iso-spin invariance. The model does not explain why $f_1 \simeq f_2$ nor why θ_C is small. But the chief defect of the model is that it contains strange neutral currents. The more complicated version of Bars, Halpern and Yoshimura (1973) avoids such currents (without introducing a charm quantum number!).

Quarks can be introduced as triplets under $SU(3)_{\text{LOCAL}}$ in (18.7) but singlets under $SU(3)_{\text{GLOBAL}}$. The connection between the strong interactions and the weak and electromagnetic ones is through the Higgs fields $K(x)$ only, since only these transform under both $U(3)$ groups in (18.7). It is surprising that the electromagnetic and weak coupling of hadrons and leptons should have the same strength, but Bars, Halpern and Yoshimura show this to be the case.

18.4 'Colour' groups

At the end of §13.5, we mentioned the conflict between the fermion character of quarks and the symmetry of the three quark states required for the lowest lying baryons. We also explained how this was overcome in the Han–Nambu model. There is another type of model which achieves the same purpose. The total group is

$$SU(3)_{\text{COLOUR}} \times SU(3)_{\text{PHYSICAL}}. \tag{18.14}$$

The difference from the Han–Nambu model is that the electric charge is independent of $SU(3)_{\text{COLOUR}}$ quantum numbers (so that quarks within an $SU(3)_{\text{COLOUR}}$ triplet are distinguished only by their 'colour'). If all known baryons are colour singlets they are totally antisymmetric under $SU(3)_{\text{COLOUR}}$, and so symmetry in the remaining quantum numbers is explained. Quarks have $\frac{1}{3}$ integral charge in the colour model, like the ordinary three quark model but unlike the Han–Nambu model (see Fritzsch, Gell-Mann and Leutwyler 1973).

If we adopt the charm scheme of chapter 9, and replace (18.14) by

$$SU(3)_{\text{COLOUR}} \times SU(4)_{\text{PHYSICAL}},$$

the γ_5 anomalies cancel between leptons and hadrons as they do in the Han–Nambu model (§13.5). This is because there are three sets of quarks each with $y = \frac{1}{3}$, as opposed to the leptons with $y = -1$.

The colour type of model, generalized if necessary, satisfies the criteria mentioned in §18.1. In general, the symmetry group has the form

$$G_W \times G_S, \qquad (18.15)$$

where G_W contains the local gauge group of the weak and electromagnetic interactions and also observed approximate symmetries of the strong interactions like $SU(3)$ (unfortunately, no-one knows what G_W is in detail). G_S is the colour group of the strong interactions, and is not directly observable. It provides the binding between quarks. The quarks transform under each of G_W and G_S. All other fields transform under one or the other but not both.

Recently it has been proposed that G_S be a local gauge group, with vector (and not axial-vector) currents. It can then be shown (Weinberg 1973c) that the absence of parity-violation to order e^2 (see §18.1) is naturally explained. First, Weinberg proves that the most general quark Lagrangian (including Yang–Mills fields) invariant under local G_S conserves parity (apparently violating terms being removable by transformations like (18.2)). Second, Weinberg considers finite corrections of order g^2 (i.e. e^2) from expressions like (16.2), (16.4). He proves that they can contribute to the quark masses (and thence, for example, to iso-spin violation) but not to parity violation.

Another merit of a local colour group is that it promises to explain the success of the quark-parton theory of high-energy inelastic lepton scattering (see §9.3(iv) and Llewellyn Smith 1972). This exciting possibility is briefly explained in the next section.

18.5 Asymptotic freedom

According to the parton model of inelastic lepton-hadron scattering at high energies and momentum transfers, the lepton interacts with a single point-like constituent of the hadron, which, for the short duration of the interaction, is approximately free. But calculations with ordinary renormalizable field theories do not support such a picture – the fields do not become free at high momentum transfers. In certain non-abelian gauge theories, however, the situation is different. We briefly describe how this can be shown.

For simplicity, take a Lagrangian with no masses or dimensional coupling constants (when $\hbar = c = 1$), like for instance a pure Yang–Mills field. To renormalize such a theory, using for example dimensional regularization (see §13.2), a dimensional parameter μ has to be

introduced. This indirectly fixes the definition of the renormalized
coupling constants $g(\mu)$ (§ 14.1). As we now show, varying μ is a useful
device for exploring high momentum-transfer behaviour of the theory

Consider a renormalized, one-particle-irreducible Green's function
for N particles of momenta $p_1, p_2, ..., p_N$. It is a function

$$G[p_1, ..., p_N; g(\mu), \mu]. \tag{18.16}$$

According to renormalization theory, G is related to a bare Green's
function G^B by an equation (with dimensions $n \neq 4$, so that everything is finite)

$$[Z(\mu)]^{-\frac{1}{2}N} G[p_1, ..., p_N; g(\mu), \mu] = G^B(p_1, ..., p_N; g^0), \tag{18.17}$$

where Z is the renormalization factor for the field concerned (see
(14.1)) and g^0 is defined in (14.36) and (14.37), for example. For
simplicity, we have chosen N identical bosons, and assumed that there
is only one coupling constant.

The right-hand side of (18.17) is independent of μ, for μ was introduced only to define the renormalized quantities. This condition may
be expressed by the Callan–Symanzik equation:

$$\left(\mu \frac{\partial}{\partial \mu} + \beta \frac{\partial}{\partial g} - \tfrac{1}{2}N\gamma\right) G = 0, \tag{18.18}$$

where
$$\beta(g, \mu) = \mu \left[\frac{\partial g}{\partial \mu}\right]_{g^0}, \tag{18.19}$$

$$\gamma(g, \mu) = \mu Z^{-1} \left[\frac{\partial Z}{\partial \mu}\right]_{g^0}. \tag{18.20}$$

If β and γ are known, (18.18) gives the μ-dependence of G. But, by a
straightforward dimensional argument,

$$G(\lambda p_1, ..., \lambda p_N; g, \mu) = \lambda^{4-N} G(p_1, ..., p_N; g, \lambda^{-1}\mu); \tag{18.21}$$

so that, from the μ-dependence of G, one can infer the behaviour as
all the 4-momenta are scaled. Clearly such scaling is incompatible
with mass-shell conditions; so the argument does not apply to S-
matrix elements. The use of (18.18) is usually confined to the Euclidean
region (where any linear combination of $p_1, ..., p_N$ is a space-like
vector), so as to avoid threshold singularities.

Let us see how to calculate β and γ in (18.19) and (18.20) (we follow
the approach of 't Hooft 1973). For the present model, with a single

parameter g, the general form (14.3) for the expansion of a renormalization constant reads

$$g^0 = \mu^{2-\frac{1}{2}n}\, \hat{g}[1 + \alpha_{11}\hat{g}^2(n-4)^{-1} + \alpha_{21}\hat{g}^4(n-4)^{-1}$$
$$+ \alpha_{22}\hat{g}^4(n-4)^{-2} + ...], \quad (18.22)$$

where $\hat{g}$ is dimensionless (see (13.9)). Differentiation of (18.22) with
respect to $\ln\mu$ (keeping g^0 fixed) yields

$$0 = \tfrac{1}{2}(n-4)\hat{g}[1 + \alpha_{11}\hat{g}^2(n-4)^{-1} + \alpha_{21}\hat{g}^4(n-4)^{-1}$$
$$+ \alpha_{22}\hat{g}^4(n-4)^{-2} + ...]$$
$$- \beta[1 + 3\alpha_{11}\hat{g}^2(n-4)^{-1} + 5\alpha_{21}\hat{g}^4(n-4)^{-1}$$
$$+ 5\alpha_{22}\hat{g}^4(n-4)^{-2} + ...]. \quad (18.23)$$

Since renormalization theory guarantees that g is finite for any choice
of μ, β must be finite (that is, devoid of poles at $n = 4$). Therefore
(18.23) can be solved, iteratively for β, giving

$$-\beta = -\tfrac{1}{2}(n-4)\hat{g} + \alpha_{11}\hat{g}^3 + 2\alpha_{21}\hat{g}^5 +, \quad (18.24)$$

$$2\alpha_{22} - 3\alpha_{11}^2 = 0, \text{ etc.} \quad (18.25)$$

The latter relation is a constraint on the α_{lm} which is a consequence of
the renormalizability of the theory (in general, all α_{lm} are determined
by the α_{11}). If $(n-4) \to 0$ in (18.24), the terms remaining result from
cancellation between the $(n-4)$ factors from differentiation of $\mu^{2-\frac{1}{2}n}$
and the $(n-4)^{-1}$ divergences. (In this respect, β is like the anomalies
discussed in §13.3.) The function γ in (18.20) can be found in a similar
way.

For free fields (or if there were no divergences) G in (18.16) would
have been independent of μ (in fact, no quantity μ would have been
necessary). The presence of β and γ in (18.18) causes departures from
free-field scaling behaviour.

As a preliminary to solving (18.18), we need to solve (18.19) to
obtain $g(\mu)$ and in particular to obtain the asymptotic form of $g(\mu)$
for large μ. If $g(\mu) \to 0$ as $\mu \to \infty$, the lowest order term in (18.24) is
adequate in (18.19), and it gives

$$[g(\mu)]^2 \sim (A + 2\alpha_{11}\ln\mu)^{-1}, \quad (18.26)$$

where A is a constant. The condition for the validity of (18.26) is

$$\alpha_{11} > 0. \quad (18.27)$$

If this is the case, the effective coupling constant for large momentum
transfers tends to zero, and it may be possible to explain why the

parton model works. (Even if (18.27) holds, the γ term in (18.18) produces logarithmic deviations from free-field scaling.)

The condition (18.27), which is a necessary one for 'asymptotic freedom', is not met in any ordinary field theory with a three-field coupling. In electrodynamics, there is a physical argument which suggests that α_{11} should be negative. First, we note that, with a high-energy cut-off Λ, the pole $(4 - n)^{-1}$ is replaced by the positive quantity $\ln (\Lambda/\mu)$. Therefore (18.27) requires

$$g > g^0. \tag{18.28}$$

But in electrodynamics charge renormalization may be interpreted as due to the polarization of the vacuum. The effect is expected to be a shielding of the bare charge, implying

$$e < e^0. \tag{18.29}$$

There are also rigorous inequalities which lead to this result. One has

$$e = Z_1^{-1} Z_2 Z_3^{\frac{1}{2}} e^0 = Z_1^{-1} Z_2 e^{\mathrm{B}}, \tag{18.30}$$

where Z_1, Z_2, Z_3 respectively are the renormalization constants for the vertex-part and the electron and photon fields. But (at least in the usual gauges) in electrodynamics

$$e = e^{\mathrm{B}} \quad \text{or} \quad Z_2^{-1} Z_1 = 1, \tag{18.31}$$

and it can be proved that (Källén 1972: 215)

$$Z_3 < 1. \tag{18.32}$$

For non-abelian gauge theories, however, these arguments fail. Equation (18.31) is not true (except possibly in the axial gauge (3.10)) because the generalized Ward identity relates the ratio g^{B}/g to spurion graphs (see §14.2, Taylor 1971, Slavnov 1972a). And Z_3 is gauge-dependent, so (18.32) may not be true either. In fact it is found, for a pure Yang–Mills theory, that

$$\alpha_{11} = (11/48\pi^2) C_2, \tag{18.33}$$

where C_2 is the constant connected with the group, defined by

$$f_{\alpha\beta\gamma} f_{\alpha'\beta\gamma} = C_2 \delta_{\alpha\alpha'} \tag{18.34}$$

(the structure constants being those in (6.27)). The contribution of fermions to α_{11} is negative, but it does not reverse the sign of (18.33) unless there are several fermions.

If scalar fields are introduced in order to make the spin-1 particles massive by Higgs' mechanism, an extra coupling constant λ is required. Equation (18.19) must then be replaced by coupled equations in g and

λ, and asymptotic freedom would require *both* g and λ to tend to zero as $\mu \to \infty$. It has not been found possible to construct a model which is asymptotically free and which has not more than one massless vector particle (Gross and Wilczek 1973).

Because of infra-red divergences, Yang–Mills theories (without Higgs fields) cannot be consistently interpreted by conventional perturbation theory (see §4.4). In fact (18.27), being the right sign for asymptotic freedom for large momentum transfers, is the *wrong* sign for freedom for small momenta (the infra-red region). It is perhaps not impossible that, if properly understood, a Yang–Mills theory might make a satisfactory basis for strong interactions.

With this hope in mind, one can work out some consequences of such an asymptotically-free theory. There are arguments that a Callan–Symanzik equation like (18.18) applies to the moments of the structure functions in inelastic lepton-hadron scattering. In a free-field theory, these moments would be constants, independent of the momentum transfer q^2. In the asymptotically-free field theory, the 'anomalous dimension' γ in (18.18) leads to a logarithmic dependence on q^2 (at values of q^2 for which $g(-q^2)$ is small). Striking effects are predicted (Gross 1974), but experiments are not yet able to detect them.

For further details on asymptotic freedom, see the review by Politzer (1974).

18.6 Weak non-leptonic decays

These decays require the calculation of matrix-elements like

$$\int \mathrm{d}^4 x \langle A \text{ in } |T(J^\lambda(x) J^\dagger_\lambda(0))| B \text{ out} \rangle D_\mathrm{F}(x; M) \qquad (18.35)$$

in which $D_\mathrm{F}(x; M)$ is the Feynman propagator of a mass M boson, A and B are hadron states, and J_λ is the hadronic part of a weak current. Because M is large, $D_\mathrm{F}(x; M)$ is heavily damped except for $x^2 \simeq 0$. The expansion of Wilson (1969) is designed to approximate operator products, like the one in (18.35), near the light-cone. The behaviour of the matrix-elements of the operators in the Wilson expansion can be studied by the methods of the Callan–Symanzik equation, outlined in the last section.

With a model of weak and strong interactions, like that of the colour model discussed in §18.4, which exhibits asymptotic freedom, one can attempt to calculate the leading terms in expressions such as (18.35).

Gaillard and Lee (1974*a*) and Altarelli and Maiani (1974) have ex-
amined a number of processes by this method. One interesting finding
is that the strong interactions provide *some* enhancement of the
$\Delta T = \frac{1}{2}$, as compared with the $\Delta T = \frac{3}{2}$, non-leptonic transitions. It
is not clear whether the enhancement is sufficient to account for the
observed accuracy of the $\Delta T = \frac{1}{2}$ rule (see, for instance, Bailin 1971).

18.7 Hierarchical symmetry-breaking

There have been a number of suggestions that perhaps an underlying
large local gauge symmetry (of all interactions) is broken down in a
succession of steps, giving a hierarchy of broken symmetries. Start
with a gauge group G and Higgs fields ϕ. Suppose that the vacuum-
expectation-value

$$F = F^{(0)} + \epsilon F^{(1)} + \epsilon^2 F^{(2)} + \dots, \qquad (18.36)$$

where ϵ is some small parameter (perhaps a power of the coupling
constant g). Let the little-group of $F^{(s)}$ be $G^{(s)}$ ($s = 0, 1, 2, \dots$), with

$$G \supset G^{(0)} \supset G^{(1)} \supset G^{(2)} \text{ etc.} \qquad (18.37)$$

Then G is strongly broken down to $G^{(0)}$, with very heavy vector mesons
generated in the process (masses of order $gF^{(0)}$), $G^{(0)}$ is less strongly
broken down to $G^{(1)}$ with less heavy vector mesons; and so on. The
observable effects of the super-heavy vector mesons might be very
small at ordinary energies.

One of the advantages of such a scheme is that G could be a simple
group with a single coupling constant g (see § 7.2), while another group
along the chain (18.37), say $G^{(1)}$, could be the Salam–Weinberg group
$SU(2) \times U(1)$ (see chapter 8). The angle θ_W would then be determined.

No one has yet found a model in which a structure like (18.36)
appears naturally. Nevertheless, we mention three examples, in each
of which the existence of hierarchical symmetry-breaking is *assumed*.

(i) *Weinberg's $SU(3) \times SU(3)$ model of leptons*

This is based on the local group $SU(3)_L \times SU(3)_R$ combined with the
discrete parity operation P (Weinberg 1971*b*). Because P destroys the
direct product structure, there is a single coupling constant g (see
§ 7.2). The leptons (μ_L^+, ν_e, e_L^-) and $(\mu_R^+, \nu_\mu^c, e_R^-)$ (where ν_μ^c is the charge-
conjugate of the ν_μ field) form $(1, 3)$ and $(3, 1)$ representations. The
Salam–Weinberg group is a subgroup. The model predicts that

$$|\theta_W| = 30^0. \qquad (18.38)$$

Parity non-conservation is a result of the first stage of symmetry-breaking.

Muon and electron lepton-numbers are not exactly conserved, and the suppression of unobserved processes like

$$\mu^- + (A, Z) \to e^+ + (A, Z - 2) \qquad (18.39)$$

(where (A, Z) is a nucleus of atomic weight A and atomic number Z) is attributed to the large mass of the super-heavy vector mesons.

(ii) *The SU(5) model of Georgi and Glashow*

This model (Georgi and Glashow 1974) is designed to include weak, electromagnetic and strong interactions in as economical way as possible. Strong interactions are assumed to be generated by an $SU(3)_{\text{COLOUR}}$ gauge group, as in §18.4. Thus

$$SU(5) \supset [SU(3)_{\text{COLOUR}}] \times [SU(2) \times U(1)_{\text{WEINBERG-SALAM}}]. \quad (18.40)$$

The 6 lepton states and the 24 quark states (counting different helicities as different states, and having three quartets of quarks) belong to two right-handed *5* and two left-handed *10* dimensional representations of $SU(5)$. The decays of baryons to leptons are suppressed by assuming very high masses for the super-heavy vector mesons.

At first sight, this model gives the Weinberg angle to be

$$\tan^2 \theta_{\text{W}} = g'^2/g^2 = \tfrac{3}{5}. \qquad (18.41)$$

Care must be taken, however, with the definition of the coupling constants: various definitions can differ by terms of order $g^2 \ln M_s$, where M_s is some super-heavy mass. According to Georgi, Quinn and Weinberg (1974), (18.41) holds for asymptotically defined coupling constants ($\mu \to \infty$ in the notation of §18.5); and the measured value of θ_{W} should be smaller than (18.41).

(iii) *The SU(4) × SU(4) model of Pati and Salam*

Like the previous model, this (Pati and Salam 1974) includes leptons and hadrons in the same multiplet. Twelve (integrally charged) quarks and four leptons are put into a $(4, \overline{4})$ representation of

$$SU(4) \times SU(4).$$

There is no absolute baryon conservation, and quarks are indeed unstable. But, since protons have 'quark-lepton number' 3, their decays involve at least three leptons, and so the proton life-time may be very long indeed.

References

Abers, E. S., Dicus, D. E., Norton, R. E. & Quinn, H. R. (1968). *Phys. Rev.* **167**, 1461.

Abers, E. S. & Lee, B. W. (1973). *Phys. Reports* **9**C, no. 1.

Adler, S. L. (1966). *Phys. Rev.* **143**, 1144.

Adler, S. L. (1974). *Phys. Rev.* D**9**, 229.

Adler, S. L. & Bardeen, W. A. (1969). *Phys. Rev.* **182**, 1517.

Adler, S. L. & Dashen, R. F. (1968). *Current Algebras and Applications to Particle Physics*. New York: Benjamin.

Adler, S. L., Nussinov, S. & Paschos, E. A. (1974). *Phys. Rev.* D**9**, 2125.

Albright, C. H., Lee, B. W., Paschos, E. A. & Wolfenstein, L. (1973). *Phys. Rev.* D**7**, 2220.

Altarelli, G. A. & Maiani, L. (1974). *Phys. Lett.* **52**B, 351.

Anderson, P. W. (1958). *Phys. Rev.* **110**, 827.

Anderson, P. W. (1959). *Phys. Rev.* **112**, 1900.

Anderson, P. W. (1963). *Phys. Rev.* **130**, 439.

Angerson, W. (1974). *Nucl. Phys.* B**69**, 493.

Appelquist, T. W., Primack, J. R. & Quinn, H. R. (1973). *Phys. Rev.* D**7**, 2998.

Ashmore, J. F. (1972). *Nuovo Cimento Lett.* **4**, 289.

Asratyan, A. E. *et al.* (1974). *Phys. Lett.* **49**B, 488.

Aubert, B. *et al.* (1974a). *Phys. Rev. Lett.* **32**, 1457.

Aubert, B. *et. al.* (1974b). *Phys. Rev. Lett.* **33**, 984.

Aviv, R. & Zee, A. (1972). *Phys. Rev.* D**5**, 2372.

Bailin, D. (1971). *Rep. Prog. Phys.* **34**, 491.

Bailin, D. & Love, A. (1974). *Nucl. Phys.* B**69**, 142.

Baker, M. & Glashow, S. L. (1962). *Phys. Rev.* **128**, 2462.

Bardeen, W. A. (1969). *Phys. Rev.* **184**, 1848.

Bardeen, J., Cooper, L. N. & Schrieffer, J. B. (1957). *Phys. Rev.* **106**, 162.

Barish, B. C. *et al.* (1974). *Phys. Rev. Lett.* **32**, 1387.

Barish, B. C. *et al.* (1975). *Phys. Rev. Lett.* **34**, 538.

Barish, S. J. *et al.* (1974). *Phys. Rev. Lett.* **33**, 448.

Bars, I., Halpern, M. B. & Yoshimura, M. (1973). *Phys. Rev.* D**7**, 1233.

Becchi, C., Rouet, A. & Stora, R. (1976). *Ann. Phys.* **98**, 287.

Bell, J. S. (1973). *Nucl. Phys.* B**60**, 427.

Benvenuti, A. *et al.* (1974). *Phys. Rev. Lett.* **32**, 800.

Benvenuti, A. *et al.* (1975). *Phys. Rev. Lett.* **34**, 419.

Bernstein, J. (1974). *Rev. Mod. Phys.* **46**, 7.

Bjorken, J. D. & Drell, S. D. (1965). *Relativistic Quantum Fields*. New York: McGraw-Hill.

Bjorken, J. D. & Llewellyn Smith, C. H. (1973). *Phys. Rev.* D**7**, 887.

Bogoliubov, N. N. & Shirkov, D. V. (1959). *Introduction to the Theory of Quantized Fields*. English translation G. M. Volkoff. New York: Wiley.

Bollini, C. G. & Giambiagi, J. J. (1972). *Phys. Lett.* **40** B, 566.

Bouchiat, M. A. & Bouchiat, C. C. (1974). *Phys. Lett.* **48** B, 111.

Boulware, D. G. (1970). *Ann. Phys.* **56**, 140.

Budny, R. (1973). *Phys. Letters.* **45** B, 340.

Budny, R. & McDonald, A. (1974). *Phys. Lett.* **48** B, 423.

Cabibbo, N. (1963). *Phys. Rev. Lett.* **10**, 531.

Capper, D. M. & Liebrandt, G. (1973). *Lett. al Nuovo Cimento*, **6**, 117.

Chaloupka, V. *et al.* (1974). *Phys. Lett.* **50** B, no. 1.

Coleman, S. & Weinberg, E. (1973). *Phys. Rev.* D **7**, 1888.

Collins, J. C. (1974), *Nucl. Phys.* B **80**, 341.

Combley, F. & Picasso, E. (1974). *Phys. Reports* **14**, no. 1.

Cornwall, J. M., Levin, D. N. & Tiktopoulos, G. (1974). *Phys. Rev.* D **10**, 1145.

Cornwall, J. M. & Norton, R. E. (1973). *Phys. Rev.* D **8**, 3338.

Cundy, D. C. *et al.* (1970). *Phys. Lett.* **31** B, 478.

Delbourgo, R. & Akyeampong, D. A. (1974). *Nuovo Cimento* **19** A, 219.

De Wit, B. (1973). *Nucl. Phys.* B **51**, 237.

De Witt, B. S. (1967). *Phys. Rev.* **162**, 1195 & 1239.

Dicus, A. D. & Mathur, V. S. (1973). *Phys. Rev.* D **7**, 525.

Dirac, P. A. M. (1948). *Phys. Rev.* **74**, 817.

Dixon, J. A. (1975). *Nucl. Phys.* to be published.

Dolan, L. & Jackiw, R. (1974). *Phys. Rev.* D **9**, 3320.

Duncan, A. & Schattner, P. (1973). *Phys. Rev.* D **7**, 1861.

Eden, R. J., Landshoff, P. V., Olive, D. I. & Polkinghorne, J. C. (1966). *The Analytic S-matrix*. Cambridge. (Particularly section 2.9.)

Eichten, T. *et al.* (1973). *Phys. Lett.* **46** B, 281.

Eliezer, S. (1974). *Phys. Lett.* **53** B, 86.

Englert, F. & Brout, R. (1964). *Phys. Rev. Lett.* **13**, 321.

Englert, F. & Brout, R. (1974). *Phys. Lett.* **49** B, 77.

Faddeev, L. D. & Popov, V. N. (1967). *Phys. Lett.* **25** B, 29.

Feynman, R. P. (1963). *Acta Phys. Polonica* **24**, 297.

Feynman, R. P. (1972). *Photon–Hadron Interactions*, Reading, Massachusetts: Benjamin.

Feynman, R. P. & Gell-Mann, M. (1958). *Phys. Rev.* **109**, 193.

Feynman, R. P. & Hibbs, A. R. (1965). *Quantum Mechanics and Path Integrals.* New York: McGraw-Hill.

Fradkin, E. S. & Tyutin I. V. (1970). *Phys. Rev.* D **2**, 2841.

Frampton, P. H. (1974). *Dual Resonance Models.* Reading, Massachusetts: Benjamin.

Freedman, D. Z. (1974). *Phys. Rev.* D **9**, 1389.

Freedman, D. Z. & Kummer, W. (1973). *Phys. Rev.* D **7**, 1829.

Fritzsch, H., Gell-Mann, M. & Leutwyler, H. (1973). *Phys. Lett.* **47** B, 365.

Gaillard, M. K. & Lee, B. W. (1974 a). *Phys. Rev. Lett.* **33**, 108.

Gaillard, M. K. & Lee, B. W. (1974 b). *Phys. Rev.* D **10**, 897.

Gasiorowicz, S. (1966). *Elementary Particle Physics.* New York: Wiley.

Gastmans, M. & Meuldermans, R. (1973). *Nucl. Phys.* B **63**, 277.

Gell-Mann, M. (1964). *Physics* **1**, 63.

Gell-Mann, M. & Lévy, M. (1960). *Nuovo Cimento* **16**, 705.

Gell-Mann, M. & Ne'eman, Y. (1964). *The Eightfold Way.* New York: Benjamin.

Georgi, H. & Glashow, S. L. (1973 a). *Phys. Rev.* D **6**, 429.

Georgi, H. & Glashow, S. L. (1973 b). *Phys. Rev.* D **7**, 2457.

Georgi, H. & Glashow, S. L. (1974). *Phys. Rev. Lett.* **32** 438.

Georgi, H. & Pais, A. (1974). *Phys. Rev.* D10, 1246.

Georgi, H., Quinn, H. R. & Weinberg, S. (1974). *Phys. Rev. Lett.* 33, 451.

Gilbert, W. (1964). *Phys. Rev. Lett.* 12, 713.

Gilmore, R. (1974). *Lie Groups, Lie Algebras and Some of their Applications.* New York: Wiley.

Glashow, S. L. (1961). *Nucl. Phys.* 22, 579.

Glashow, S. L., Iliopoulos, J. & Maiani, L. (1970). *Phys. Rev.* D2, 185.

Goldstone, J. (1961). *Nuovo Cimento* 19, 154.

Goldstone, J., Salam, A. & Weinberg, S. (1962). *Phys. Rev.* 127, 965.

Good, R. H. (1955). *Rev. Mod. Phys.* 27, 187.

Gross, D. J. (1974). *Phys. Rev. Lett.* 32, 1071.

Gross, D. J. & Jackiw, R. (1972). *Phys. Rev.* D6, 477.

Gross, D. J. & Wilczek, F. (1973). *Phys. Rev.* D8, 3633.

Guralnik, G. S., Hagen, C. R. & Kibble, T. W. B. (1964). *Phys. Rev. Lett.* 13, 585.

Guralnik, G. S., Hagen, C. R. & Kibble, T. W. B. (1968). *Advances in Particle Physics* 2, 567.

Gurr, M. S., Reines, F. & Sobel, H. W. (1974). *Phys. Rev. Lett.* 33, 179.

Hagiwara, T. & Lee, B. W. (1973). *Phys. Rev.* D7, 459.

Hamermesh, M. (1960). *Group theory and its applications to physical problems.* New York: Addison-Wesley.

Han, M-Y. & Nambu, Y. (1965). *Phys. Rev.* 139, B1006.

Harrington, B. J. & Yildiz, A. (1974). *Phys. Rev. Lett.* 33, 324.

Hasert, F. J. *et al.* (1973a). *Phys. Lett.* 46B, 121.

Hasert, F. J. *et al.* (1973b). *Phys. Lett.* 46B, 138.

Higgs, P. W. (1956). *Nuovo Cimento* 4, 1262.

Higgs, P. W. (1964a). *Phys. Lett.* 12, 132.

Higgs, P. W. (1964b). *Phys. Rev. Lett.* 13, 508.

Higgs, P. W. (1966). *Phys. Rev.* 145 1156.

Honerkamp, J. (1972). *Nucl. Phys.* B48, 269.

Jackiw, R. (1972). In *Lectures on Current Algebra and its applications* by Treiman, S. B., Jackiw R. & Gross D. J., Princeton: University Press.

Jackiw, R. (1974). *Phys. Rev.* D9, 1686.

Jackiw, R. & Johnson, K. (1973). *Phys. Rev.* D8, 2386.

Jackiw, R. & Weinberg, S. (1972). *Phys. Rev.* D5, 2396.

Jacobs, L. (1974). *Phys. Rev.* D10, 3956.

Jauch, J. M. & Rohrlich, F. (1955). *The Theory of Photons and Electrons.* New York: Addison-Wesley.

Jones, D. R. T. (1974). *Nucl. Phys.* B71, 111.

Källén, G. (1972). *Quantum Electrodynamics*, translated C. K. Iddings & M. Mizushima. London: Allen & Unwin.

Kamefuchi, S. (1960). *Nucl. Phys.* 18, 691.

Keller, J. B. & Zumino, B. (1961). *Phys. Rev. Lett.* 7, 164.

Kibble, T. W. B. (1967). *Phys. Rev.* 155, 1554.

Kirzhnits, D. A. & Linde, A. D. (1972). *Phys. Lett.* 42B, 471.

Kluberg-Stern, H. & Zuber, J. B. (1975a). 'Renormalization of non-abelian gauge theories in a background field gauge: I Green functions', *Phys. Rev.* D, to be published.

Kluberg-Stern, H. & Zuber, J. B. (1975b). 'Renormalization of non-abelian gauge theories in a background field gauge: II Gauge invariant operators', *Phys. Rev.* D, to be published.

Kobsarev, I. Yu., Okun, L. B. & Zeldovich, Ya. B. (1974). *Phys. Letters* 50B, 340.

Korthals Altes, C. P. & Perrottet, M. (1972). *Phys. Lett.* **39**B, 546.

Lange, R. (1966). *Phys. Rev.* **146**, 301.

Lee, B. W. (1972). *Phys. Rev.* D**5**, 823.

Lee, B. W. (1973). *Phys. Lett.* B**46**, 214.

Lee, B. W., Primack, J. R. & Treiman, S. B. (1973). *Phys. Rev.* D**7**, 501.

Lee, B. W. & Treiman, S. B. (1973). *Phys. Rev.* D**7**, 1211.

Lee, B. W. & Zinn-Justin, J. (1972). *Phys. Rev.* D**5**, 3137.

Lee, B. W. & Zinn-Justin, J. (1973). *Phys. Rev.* D**7**, 1049.

Lee, T. D. (1971). *Phys. Rev. Lett.* **26**, 801.

Lee, T. D. (1974). *Phys. Reports* **9**C, no. 2.

Lee, T. D. & Nauenberg, M. (1964). *Phys. Rev.* **133**, B1549.

Lee, W. (1972). *Phys. Lett.* **40**B, 423.

Lehman, H. & Pohlmeyer, K. (1971). *Comm. Math. Phys.* **20**, 101.

Li, L-F. (1974). *Phys. Rev.* D**9**, 1723.

Llewellyn Smith, C. H. (1972). *Phys. Reports* **3**C, no. 5.

Llewellyn Smith, C. H. (1973). *Phys. Lett.* **46**B 233.

Mandelstam, S. (1968). *Phys. Rev.* **175**, 1580, 1604.

Mandelstam, S. (1975). *Phys. Letters* **53**B, 476.

Marshak, R. E., Riazuddin & Ryan, C. P. (1969). *Theory of Weak Interactions in Particle Physics.* New York: Wiley.

Marshak, R. E. & Sudarshan, E. C. G. (1958). *Phys. Rev.* **109**, 1860.

Matthews, P. T. (1949). *Phys. Rev.* **76**, 1254.

Matthews, P. T. & Salam, A. (1955). *Nuovo Cimento* **2**, 120.

Michel, L. & Radicati, L. A. (1971). *Ann. Phys. (USA)* **66**, 758.

Mohapatra, R. N. (1971). *Phys. Rev.* D**4**, 378 & 1007.

Mohapatra, R. N. (1972). *Phys. Rev.* D**6**, 2023.

Mohapatra, R. N. (1974). *Phys. Rev.* D**9**, 3461.

Mohapatra, R. N. & Pati, J. C. (1975). *Phys. Rev.* D**11**, 566.

Mohapatra, R. N. & Sakakibara, S. (1974). *Phys. Rev.* D**9**, 429.

Nambu, Y. (1960). *Phys. Rev. Lett.* **4**, 380.

Nambu, Y. (1974). *Phys. Rev.* D**10**, 4262.

Nambu, Y. & Han, M.-Y. (1974). *Phys. Rev.* D**10**, 674.

Nambu, Y. & Jona-Lasinio, G. (1961). *Phys. Rev.* **122**, 345.

Nielsen, H. B. & Olesen, P. (1973). *Nucl. Phys.* B**61**, 45.

Nouri-Moghadam, M. & Taylor, J. G. (1975). *J. Phys.* A**8**, 334.

Omnès, R. (1970). *Introduction to Particle Physics,* translated by G. Barton. New York: Wiley.

Pais, A. (1973). *Phys. Rev.* D**8**, 625.

Pais, A. & Treiman, S. B. (1974). *Phys. Rev.* D**9**, 1459.

Palmer, W. F. (1972). *Phys. Rev.* D**6**, 1190.

Paschos, E. A. & Wolfenstein, L. (1973). *Phys. Rev.* D**7**, 91.

Pati, J. C. & Salam, A. (1974). *Phys. Rev.* D**10**, 275.

Perkins, D. H. (1974). In *Proceedings of the fifth Hawaii topical conference on particle physics* (1973), eds. P. N. Dobson, V. Z. Peterson & S. F. Tuan, University Press of Hawaii.

Politzer, H. D. (1974). *Phys. Reports* **14**C, 129.

Polkinghorne, J. C. (1955). *Proc. Roy. Soc.* A **230**, 272.

Preparata, G. & Weisberger, W. I. (1968). *Phys. Rev.* **175**, 1965.

Rose-Innes, A. C. & Rhoderick, E. H. (1969). *Introduction to Superconductivity.* New York: Pergamon.

Ross, D. A. (1973). *Nucl. Phys.* B**51**, 116.

Ross, D. A. & Taylor, J. C. (1973). *Nucl. Phys.* B**51**, 125.

Salam, A. (1960). *Nucl. Phys.* **18**, 681.
Salam, A. (1968). In *Elementary Particle Theory*, ed. N. Svartholm. Stockholm: Almqvist, Forlag AB, page 367.
Salam, A. & Strathdee, J. (1974). *Nature*, **252**, 569.
Salam, A. & Ward, J. C. (1964). *Phys. Lett.* **13**, 168.
Schechter, J. & Ueda, Y. (1973). *Phys. Rev.* D **7**, 3119.
Schwarz, J. H. (1973). *Phys. Reports* **8**C, no. 4.
Schwinger, J. (1957). *Ann. Phys. (N.Y.)* **2**, 407.
Schwinger, J. (1962a). *Phys. Rev.* **125**, 397.
Schwinger, J. (1962b). *Phys. Rev.* **128**, 2425.
Schwinger, J. (1966). *Phys. Rev.* **144**, 1087.
Sehgal, L. M. (1969). *Phys. Rev.* **183**, 1511.
Sehgal, L. M. (1974). *Phys. Lett.* **48**B, 60.
Shaw, R. (1955). *The Problem of Particle Types and Other Contributions to the Theory of Elementary Particles*. Cambridge Ph.D. thesis, unpublished.
Sirlin, A. (1974). *Phys. Rev. Lett.* **32**, 966.
Slavnov, A. (1972a). *Theor. & Math. Phys.* **10**, 99.
Slavnov, A. (1972b). *Theor. & Math. Phys.* **13**, 174.
Snow, G. A. (1973). *Nucl. Phys.* B **55**, 445.
Stodolsky, L. (1974). *Phys. Lett.* **50**B, 352.
Taylor, J. C. (1971). *Nucl. Phys.* B **33**, 436.
't Hooft, G. (1971a). *Nucl. Phys.* B **35**, 167.
't Hooft, G. (1971b). *Phys. Lett.* **37**B, 195.
't Hooft, G. (1973). *Nucl. Phys.* B **61**, 455.
't Hooft, G. (1974). *Nucl. Phys.* B **79**, 276.
't Hooft, G. & Veltman, M. (1972a). *Nucl. Phys.* B **44**, 189.
't Hooft, G. & Veltman, M. (1972b). *Nucl. Phys.* B **50**, 318.
Tomboulis, E. (1973). *Phys. Rev.* D **8**, 2736.
Veltman, M. (1968). *Nucl. Phys.* B **7**, 637.
Vinen, W. F. (1968). *Reports on Progress in Phys.* **31**, 61.
Weinberg, S. (1965a). *Phys. Rev.* **138**, B **988**.
Weinberg, S. (1965b). *Phys. Rev.* **140**, B **516**.
Weinberg, S. (1967). *Phys. Rev. Lett.* **19**, 1264.
Weinberg, S. (1971a). *Phys. Rev. Lett.* **27**, 1688.
Weinberg, S. (1971b). *Phys. Rev.* D **5**, 1962.
Weinberg, S. (1972). *Gravitation and Cosmology: Principles and Applications of the General Theory of Relativity*. New York: Wiley.
Weinberg, S. (1973a). *Phys. Rev.* D **7**, 1068.
Weinberg, S. (1973b). *Phys. Rev.* D **7**, 2887.
Weinberg, S. (1973c). *Phys. Rev.* D **8**, 4482.
Weinberg, S. (1973d). *Phys. Rev. Lett.* **31**, 494.
Weinberg, S. (1974). *Phys. Rev.* D **9**, 3357.
Weinstein, M. (1973a). *Phys. Rev.* D **7**, 1854.
Weinstein, M. (1973b). *Phys. Rev.* D **8**, 2511.
Wess, J. & Zumino, B. (1971). *Phys. Lett.* **37**B, 95.
Wess, J. & Zumino, B. (1974). *Phys. Lett.* **49**B, 52.
Wilson, K. (1969). *Phys. Rev.* **179**, 1499.
Wolfenstein, L. (1964). *Phys. Rev. Lett.* **13**, 562.
Yang, C. N. & Mills, R. L. (1954). *Phys. Rev.* **96**, 191.
Zee, A. (1974). *Phys. Rev.* D **9**, 1772.
Zinn–Justin, J. (1975). In *Lecture Notes in Physics 37*, eds. J. Ehlers et. al. Berlin: Springer–Verlag.

Index